Pirates of the Cell

The Story of Viruses from Molecule to Microbe

Pirates of the Cell

The Story of Viruses from Molecule to Microbe

Andrew Scott

BASIL BLACKWELL

First published 1985

Basil Blackwell Ltd
108 Cowley Road, Oxford OX4 1JF, UK

Basil Blackwell Inc.
432 Park Avenue South, Suite 1505,
New York, NY 10016, USA

British Library Cataloguing in Publication Data

Scott, Andrew
1. Viruses
I. Title
576'.64 QR360

ISBN 0-631-14046-8

Library of Congress Cataloging in Publication Data

Scott, Andrew.
Pirates of the cell.

Bibliography: p.
Includes index.
1. Virology — Popular works. 2. Viruses. I. Title.
[DNLM: 1. Virology — popular works. QW 160 S425p]
QR364.S38 1985 616'.0194 85-7347

ISBN 0-631-14046-8

Typeset by Oxford Publishing Services, Oxford.
Printed in Great Britain by Billing and Sons Ltd., Worcester

For Margaret

Contents

Preface ix
Introduction x
1 Discovery — the mysterious fluid 1
2 Building-blocks – of viruses, mice and men 12
3 Structure — the virus revealed 33
4 Multiplication — life at the limit 43
5 Integration — the virus "lies low" 61
6 Invasion — strategies for a gatecrasher 75
7 Defence — the body fights back 85
8 Damage — the virus rampant 97
9 Disease — an infinity of courses 109
10 Cancer — one culprit of many 132
11 Vaccines — priming the trap 159
12 Cures — the dawn of an era? 175
13 Mysteries — many puzzles remain 188
14 Exploitation — from menace to boon? 217
References 230
Further Reading 240
Index 241

Preface

This book is intended to introduce the viruses to anyone who is interested. It will hopefully be of use to undergraduates in medicine and the life sciences, as well as non-virologist scientists; but it is also designed to be accessible to laymen with an interest in biology.

My aim has been to try to convey all the really important facts about the viruses, without getting bogged down in unnecessary detail. I haven't shied away from important complexities, but the emphasis is always on the *general* features of viruses, the things they do, and the ways in which we can combat and use them. The bias is also very much in favour of the viruses that infect animals and especially man, since these viruses seem to be the most interesting and relevant types to use as examples of viruses in general.

Chapter 2 may be unnecessary for readers already aware of the basic biochemical logic of living things – in other words how genes make RNA, which makes proteins, which make us all work. But it will be vital and possibly quite difficult for lay readers. Actually, chapters 2, 4, 5 and 7 might all be a challenge to such readers, but if you persist through these the others should be considerably easier, and you will end up knowing more about the viruses than you ever thought possible!

I would like to thank John Davey, Sean Magee, Kim Pickin, Carol Busia and Peter Whatley for their respective ideas and efforts at Blackwell's. Thanks are also due to Dick Wall for his advice and useful discussions; and to Margaret, for all her help.

Introduction

Viruses are tiny and incredibly simple biological "pirates". They invade living cells and then exploit the complex metabolic machinery of the cells to make many more viruses. The simplicity of virus structure prompts many people to suggest that they cannot really be "alive", but are instead "mere chemicals" with the ability to multiply within truly "living" creatures such as ourselves. But if we do consider the viruses to be alive, then they clearly represent life at its simplest and most basic level.

Overall, viruses do little more than enter cells, multiply, and then escape from the cells to enter new ones and start the whole business of multiplication all over again. Almost all of the chemical reactions required to bring about this endless cycle are performed by the cells they invade, rather than by the viruses themselves. Despite the simplicity of viruses and the viral life-cycle overall, however, the process of viral infection can follow an astonishing variety of different paths; and rather than always being committed to rampant multiplication, viruses can often quietly take up residence within infected cells for long periods of time.

During the course of their wanderings from cell to cell the viruses can do a great deal of damage. The diseases they cause range from trivial infections such as the common cold to such deadly illnesses as smallpox, rabies, yellow fever and cancer. In between these two extremes come many other conditions such as influenza, chickenpox, measles, mumps, hepatitis, polio and many more – all capable of killing, maiming, or at least causing us considerable discomfort for days, weeks or months on end. Of course the viruses don't just attack humans, but can also infect other animals and plants as well as much simpler creatures such as bacteria.

This is a book about how viruses work, what they can do to us and what we can do to and with them.

I will be telling you about how we first discovered the viruses, what they are made of and what they "look" like; about how they manage to invade the vulnerable cells within our bodies, multiply within them, and then escape to spread an infection elsewhere.

We will be looking at the marvellous and complex defence systems used by our bodies to fight off the threat of viral invasion; and then considering what happens when these defences fail, allowing the viruses to triumph, at least for a while, and perhaps cause serious disease.

Next, you will learn about the simple but highly effective tricks of vaccination, that allow us to protect ourselves against specific viruses by fooling our bodies into thinking that they have battled with the viruses concerned before. Then we will look at the research that is slowly beginning to provide effective drugs that can be used to combat the viruses once an infection has really taken hold.

There are many mysteries in the world of virology, and in a separate chapter I will tell you about some of the most pressing and intriguing current puzzles: examining unexplained diseases that *might* be caused by the viruses; investigating the possibility that novel types of infectious agents, a bit like the viruses only different, might exist; and considering alternative explanations for the most impenetrable mystery of them all – namely, where have the viruses come from?

Finally, we will be looking at the increasingly successful attempts to turn the viruses from infectious agents that only *harm* us, into useful biological systems that we can *exploit* to serve our own ends.

The attempt to find good uses for the viruses is just one aspect of a remarkable transformation taking place in mankind's relationship with these age-old foes. Over thousands of years the viruses have killed many millions of humans, as well as causing great damage to our livestock and crops; but now many viral infections can be completely avoided thanks to vaccination, and an era of successful anti-viral drugs might well be on the way. The viruses have certainly not yet been "conquered" by modern man, and they will doubtless continue to kill and maim for quite some years to come, but the tide has definitely been turned.

The most dramatic evidence of our rapidly changing relationship with the viruses is the victory over smallpox – one of the most devastating infectious diseases of all, but which has now been eradicated thanks to a monumental effort of vaccination and isolation by the World Health Organisation. Over the next 30 years various other dangerous viruses might meet the same fate as smallpox virus, while the damaging effects of those that aren't eradicated might be drastically reduced.

The great change being wrought in mankind's relationship with the viruses is taking place incredibly quickly. To appreciate just how quickly, we need only consider the point at which our story begins; for to discover the origins of our knowledge about the physical, chemical and biological nature of the viruses, we need only look back as far as the closing years of the last century. The discovery of the viruses, less than 100 years ago, is the subject of chapter 1.

CHAPTER 1

Discovery – the mysterious fluid

During the closing years of the last century a young Russian botanist called Dimitri Ivanovski was investigating an infectious disease of tobacco plants known as tobacco mosaic disease. The name comes from the mosaic pattern of dark and light green patches that appears on infected leaves. Ivanovski crushed some diseased tobacco plants in a dish and collected the sap released by squeezing it out through linen. As you might expect, he was then able to transmit the disease to healthy plants by treating them with the infected sap. Next, Ivanovski passed some of the sap through a porcelain filter which was believed to be fine enough to hold back all types of micro-organisms. Surprisingly, he found that the filtered sap was still infectious (see figure 1.1). Although he didn't realise it, Ivanovski had discovered the viruses.

He was working more than 200 years after Antony van Leeuwenhoek had first used his primitive microscopes to observe the microbial world. Leeuwenhoek described the tiny creatures he saw as "animalcules", but we know them now as bacteria, protozoa, algae and fungi. In the decades directly preceding Ivanovski's investigations the work of men such as Louis Pasteur and Robert Koch had led to acceptance of the germ theory of disease, which attributed all infectious diseases to Leeuwenhoek's animalcules, especially the bacteria. So it was natural for Ivanovski to assume that the mosaic disease of his tobacco plants was due to a bacterial infection. Unfortunately he clung to this belief despite his unusual discovery. To explain the infectivity of filtered sap he simply suggested that the bacteria causing the disease produced a soluble toxin that could pass through the filter. Alternatively, he

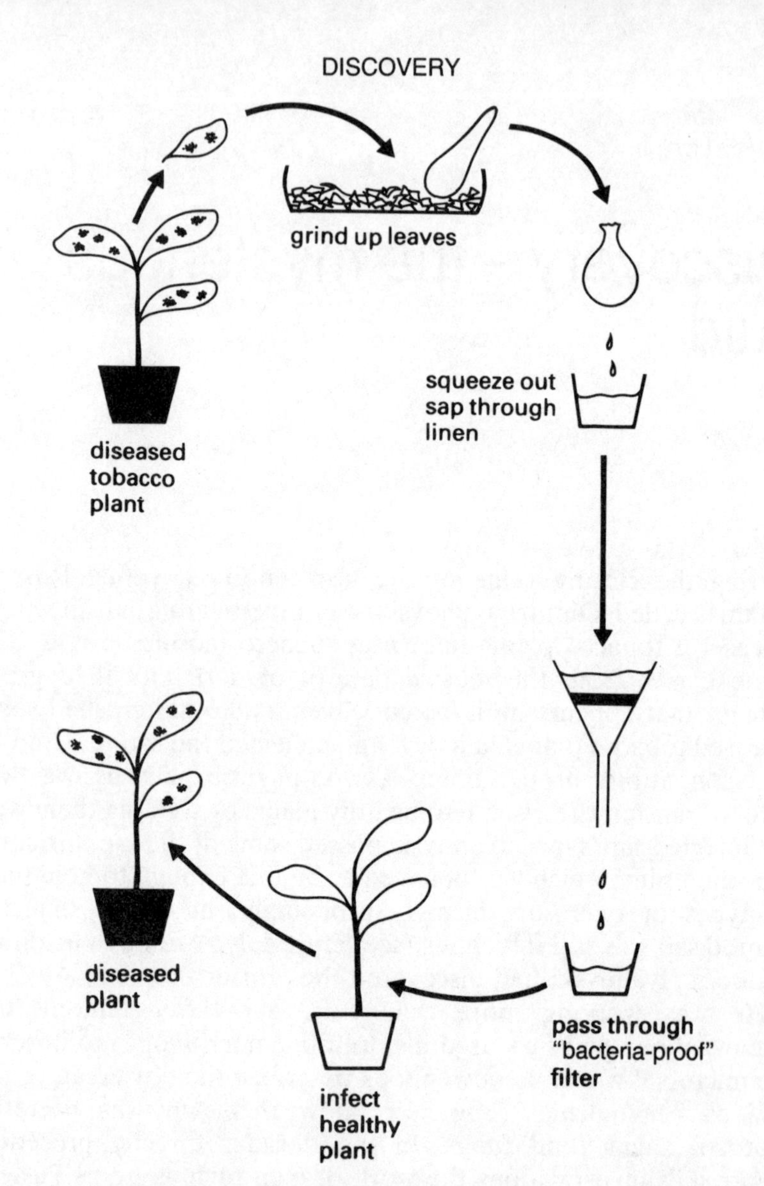

Figure 1.1 The experiment of Ivanovski and Beijerinck which led to the discovery of tobacco mosaic virus

suggested that the bacteria themselves might be unusually small and so able to pass through the filter.

Ivanovski published these conclusions in 1892, but they caused little stir in the world of microbiology. He is one of the many thousands of scientists whose claims to fame have been diminished by a failure to realise the true meaning of their results. For if he had pursued his investigations a little farther, or started with an open mind about the cause of tobacco mosaic disease, then he might be remembered as the sole and undisputed founder of the science of virology. Instead, the picture is more clouded, with credit for the discovery of the viruses sometimes given to Martinus Beijerinck – a Dutch microbiologist who performed the same experiment as Ivanovski but came to very different conclusions.

Beijerinck began a series of experiments on tobacco mosaic disease in 1897, apparently in ignorance of Ivanovski's earlier findings. Like Ivanovski, he soon found that passing infected sap through a supposedly "bacteria-proof" filter did not prevent the sap from infecting other plants. He also took a look at the sap under the microscope, but found no micro-organisms that could cause the disease. His next crucial experiment was to try to cultivate the infectious agent. He attempted to do this using methods that should have allowed bacteria to multiply, adding infected sap to watery mixtures of nutrients and keeping the culture dishes warm; but consistently he found that no micro-organism capable of causing the mosaic disease would grow.

The idea that a bacterial toxin might be passing through the filter did not appeal to Beijerinck. He had found that the filtered sap could transmit the disease successively through an "unlimited" number of plants, which implied that the agent causing the disease could multiply within the plants. So whatever was passing through the filter seemed to be some kind of living (or at least reproducing) organism, rather than a chemical toxin. There were a number of other strange features that made Beijerinck suspect he was dealing with a completely new type of organism, rather than simply a small bacterium. For example, it appeared to be resistant to alcohol, weak formalin solution and severe drying – characteristics perhaps expected of chemicals but certainly not of living organisms. Even more intriguing to Beijerinck was the ability of the infectious agent to diffuse through agar gel (a carbohydrate jelly that is mixed with nutrients to provide a medium for bacterial cultures to grow on). Only liquids and soluble substances were believed to be able to diffuse through these gels, and certainly not solid or particulate material such as living cells.

Beijerinck was thus faced with a confusing dilemma: he had an apparently living organism that passed through filters small enough to trap all known forms of life, and which often behaved more like a chemical compound than a micro-organism. In response to this confusion Beijerinck turned his back on pre-existing knowledge and took a step into the unknown. He proposed that tobacco mosaic disease was caused by a novel type of infectious agent that existed in the *fluid* or *soluble* form, rather than being made out of cells. He called this mysterious new form of life "*contagium vivum fluidum*" (soluble living germ).[1]

In making this bold and unusual suggestion Beijerinck was disregarding the widely accepted dogma that all life must be formed from single or multiple cells. This willingness to propose non-cellular forms of life is held by many people to be the speculative leap that first set the science of virology on its way, for the "soluble living germ" that caused tobacco mosaic disease was in reality a virus.

At this stage it will be worthwhile to point out that the word "virus" has undergone many changes in meaning throughout its long history.[2] In Beijerinck's day it was actually used to describe all types of infectious agent, including bacteria, protozoa and so on. With time, however, it gradually came to be restricted to only those strange filter-passing agents first discovered (although he did not realise it) by Ivanovski. To avoid confusion I will always use the word in this modern form, even when describing work performed before the modern usage was adopted.

Beijerinck's revolutionary ideas about the nature of viruses do not agree precisely with the picture later revealed by modern science. But his work did uncover one of the central features of the life-cycle of all viruses: despite the fact that it could not be cultivated artificially, in nutrient mixtures, the tobacco mosaic virus was found to multiply with no trouble at all inside infected plants. This made Beijerinck decide that in order to reproduce itself his "soluble living germ" must become "incorporated into the living protoplasm of the cell". With this inspired statement he summarised what we now recognise to be the most distinctive characteristic of the viruses: outside of living cells they are inert and "lifeless". Only when they become incorporated into the metabolism of a "host" cell can they display the properties such as reproduction that are normally associated with life. This is why the viruses are often called "obligate parasites", since they are *obliged* to parasitise living cells in order to multiply.

You need only consider the work of Ivanovski and Beijerinck to appreciate how important the *interpretation* of results is to progress in science. They both performed similar experiments and got similar results, but their conclusions could hardly have been more different; and these differing conclusions would have in turn set future research off in entirely different directions. Their work also illustrates the frequent need for several independent attempts to be made before a problem is eventually solved. In fact Ivanovski and Beijerinck were not the only people grappling with the viruses around the turn of the century. Two German scientists also found the infectious agent responsible for foot-and-mouth disease to have strange properties similar to those of tobacco mosaic virus.

Friedrich Loeffler and Paul Frosch had been asked by the German government to look into the problem of repeated and very damaging outbreaks of foot-and-mouth disease in German cattle. Their results eventually demonstrated that viruses can infect animals as well as plants.[3] They began by collecting fluid (lymph) from the vesicles that form in the mouth and udders of infected cattle. They added the lymph to nutrient solutions but were unable to get the foot-and-mouth agent to grow. Next, they tried filtering the infected lymph through "bacteria-proof" filters. After the filtration the lymph could still infect healthy animals, just as the filtered sap from diseased tobacco plants could still cause tobacco mosaic disease. Like Ivanovsky, Loeffler and Frosch suggested that bacteria might be making a soluble toxin that could pass through the filter. They also considered the idea that the organism causing the disease might itself pass through the filter, but unlike Ivanovski they did not insist that it must be a bacterium. The possibility that an unknown type of organism might be involved can be dimly perceived in their description of the infectious agent as "an agent capable of reproducing . . . so small that the pores of a filter which will hold back the smallest bacterium will still allow it to pass".

Ivanovski, Beijerinck and Loeffler and Frosch provided a set of simple criteria that could be used to identify the filter-passing agents we now call the viruses. These initial criteria were all negative observations: the viruses could not be seen through the microscope, they could not be cultivated unless introduced into another living organism, and of course they were not held back by filters that prevented the passage of bacteria and other known micro-organisms.

In subsequent years a great many infectious agents were shown to meet these criteria, allowing them to be classified as viruses long before anyone had any clear idea of what a virus really is. The diseases found to be caused by viruses in these early days included such medically and economically important conditions as rabies, poliomyelitis, measles, sheep pox and cattle plague.

In 1915 the known range of the viruses was expanded even farther when Frederick Twort, a British bacteriologist, demonstrated that they could also attack bacteria.[4] He had been growing bacterial cells on culture "plates" containing agar gel mixed, as usual, with simple nutrients required for bacterial growth. Colonies of bacteria grown in this way eventually become visible spots containing millions of individual bacterial cells. But Twort noticed that some of the spots had an uncharacteristic glassy appearance. He also found that samples taken from these glassy colonies would not multiply into fresh colonies when transferred to other agar plates. The important clue to what was wrong with the glassy colonies came when he passed material taken from these colonies through "bacteria-proof" filters. It should come as no surprise to you by now that the liquid coming through the filter could cause normal bacterial colonies to take on the glassy appearance. This indicated that a virus was causing the bacteria to form the glassy spots. We now know that the viruses were multiplying inside the bacterial cells, eventually causing the cells to burst open and release a crop of newly formed virus particles ("virus particle" is the name given to a single individual virus). It was the bursting open of the bacteria that produced the glassy appearance, and obviously samples of the burst cells would not be able to multiply on fresh agar, just as Twort had found.

As happens so often in science, Twort's important discovery attracted little attention. In 1917, however, Felix d'Herelle of the Pasteur Institute in Paris discovered the same phenomenon.[5] This time the discovery aroused great interest and was widely discussed. d'Herelle coined the term "bacteriophage" (bacteria-eater) for those viruses that infect bacteria, and the bacteriophages have since become some of the most intensively studied organisms on the Earth. Studying them has provided a wealth of information, not only about viruses, but also about the basic molecular processes of all forms of life.

Of course the discovery of viruses that can destroy bacteria also raised the appealing possibility of using these viruses to treat bacterial infections. Twort, in particular, was attracted to this notion, and he devoted considerable time and effort to the quest for bacteriophages that could be used to treat infection. These endeavours came to an abrupt end when his laboratory was destroyed by a German bomb during World War II, and the idea was largely forgotten. It has recently resurfaced, however, and as we shall see in chapter 14 the use of bacteriophages to fight disease once more looks like a promising idea.

A closer look

As more and more viral infections were discovered (based on their ability to satisfy the criteria outlined above) the debate about what type of organisms the viruses really were heated up. Beijerinck's idea of a mysterious fluid made up of "soluble living germs" was certainly not accepted by everyone. Many microbiologists felt that the viruses would simply turn out to be very small micro-organisms with the same general cellular structure as other forms of life. It was obviously a very unsatisfactory state of affairs to know of a whole series of infections caused by these agents, and yet remain completely in the dark about what they were made of or "looked like".

One of the most obvious things to do was to try to work out the *size* of viruses more precisely, always assuming that they could be described as having "size". Around the 1930s a closer look at the size of viruses became possible, thanks to the development of very fine filters with pores of known diameter. William Elford, of the National Institute for Medical Research in London, used a material called "collodion" (a modified form of cellulose) to construct a range of membrane filters with different sizes of pores. By passing infected fluids through these membranes and then testing to see if what came through was still infectious, Elford was able to estimate the size of several viruses.[6] His experiments produced two very important results. Firstly, the viruses were shown to be particulate entities which did indeed have a definite size. For foot-and-mouth disease virus, for example, he came up with a size of around 10 nanometres (one nanometre is one thousand millionth of a metre). Secondly, Elford found that viruses causing different diseases were different sizes, although those causing any one disease all seemed to be the same size. The poxviruses for example, which cause smallpox, cowpox and so on, were apparently several times larger than foot-and-mouth disease virus.

Elford's figures for viral size were startling because they were many times smaller than the dimensions accepted for the smallest cells. They prompted some scientists to doubt that anything so small could really be alive. The debate about the viruses' entitlement to be classed as *living* organisms continues to some extent right up to the present day, as I shall discuss later.

Modern techniques such as electron microscopy have revealed that foot-and-mouth disease virus is actually about twice the size originally proposed by Elford, but this is still vanishingly small. The size of different viruses actually varies between about 10 and 300 nanometres along their longest axis (see figure 1.2). There also seems to be no clear-cut distinction between the size of viruses and bacteria. Although it is certainly true that viruses are much smaller than most bacteria, the largest viruses are about the same size as the smallest bacteria.

Although Elford's estimates of viral size were not particularly accurate his work did at last demonstrate that the viruses are concrete physical entities. The fact that very fine filters would hold back the viruses was of course a great blow to Beijerinck's idea of an infectious fluid which would pass through even the finest of filters. By that time, however, Beijerinck's revolutionary concept had done its job, opening up people's minds to the idea that the viruses might be a new and completely different form of life. His belief that the viruses were non-cellular agents that could multiply only *within* living cells has since been completely confirmed.

So Elford's experiments began the process of properly defining what a virus really is; but having gained some idea of viral size, everyone was obviously very keen to find out what they were made of, how they were able to multiply and so on. Since it was impossible to cultivate the viruses by simply adding them to mixtures of chemical nutrients it was difficult to obtain large quantities of pure viruses for study. The only way to obtain a reasonable number of virus particles was to infect some suitable cells and allow the virus to multiply inside them. It was quite easy to get samples of bacteriophages in this way, because the infected bacterial cells could be easily cultivated, but working with animal or plant viruses was more difficult. In the early days whole animals or plants had to be infected and the infected tissues later ground up with sand or glass to burst open the cells and release the crop of viruses. This was all very tedious, and tending

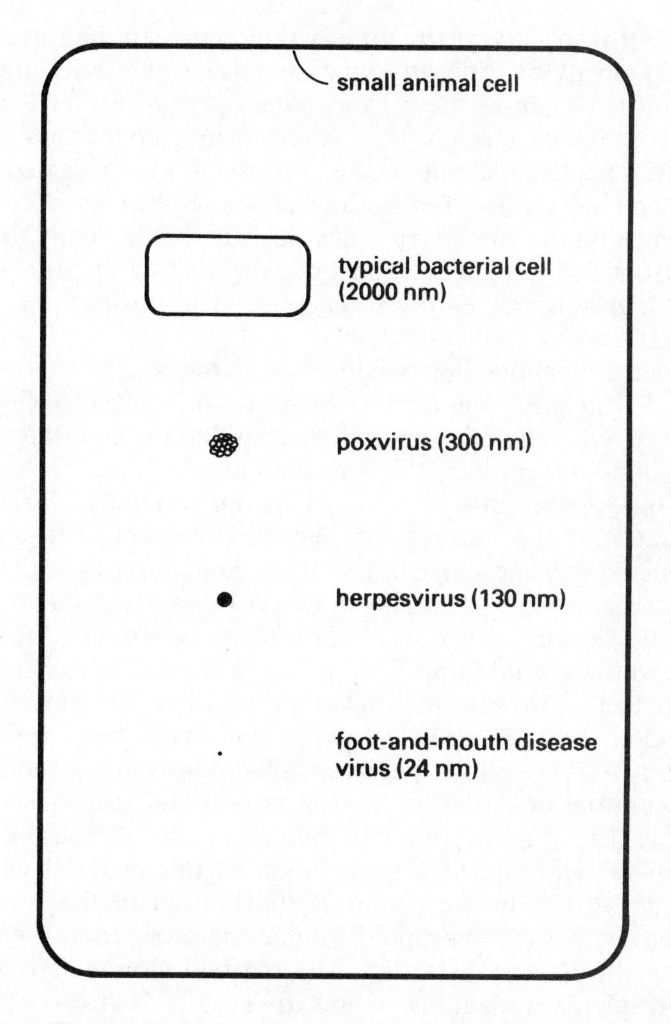

Figure 1.2 The relative sizes of some viruses and cells (1 nanometre (nm) = 1×10^{-9} metres)

the stocks of animals and plants consumed considerable time and money. Fortunately, methods have since been devised to grow many types of isolated animal cells by a technique called cell or tissue "culture". This allows a layer of cells to be grown in a culture dish, providing a cheap and convenient environment in which to grow up viruses. More recently, tissue culture methods have also been applied to plant cells, again providing a more convenient way of propagating and studying the plant viruses.

Of course to harvest the viruses they must still be released by bursting open the cells in some way (in some cases the viral infection itself causes the cells to burst open, as we have already noted with the bacteriophages). After opening up the cells a messy mixture remains, made up of the viruses and the unwanted debris from their host cells. Before any meaningful analysis could be performed on the viruses ways obviously had to be found to purify them from this mixture. Throughout the 1930s and 1940s several research groups tackled the problem of virus purification using a technique called "centrifugation".

In centrifugation the mixture of viruses and cell debris, suspended in some solution, is spun around extremely fast. The tube containing the mixture is aligned so that the centrifugal force many thousands of times stronger than gravity forces the contents of the tube downwards. But everything does not move down at the same rate; instead, the rate of downward movement depends on the density, size and shape of the different materials present in the mixture. So if the speed and duration of the centrifugation are carefully chosen you can stop the process when some materials have formed a solid lump (the pellet) in the bottom of the tube, while others are still suspended in solution. In this way the suspended material and the pellet will have been separated. Normally a series of low-speed centrifugations will send cell debris to the bottom of the tube, leaving viruses still suspended in the solution. This fluid portion can then be drawn off and put into a clean tube. High-speed centrifugation of this material will then push the viruses into a clump at the bottom of the tube. The solution, containing unwanted soluble materials from the cell, is then drawn off and discarded. The purified viruses may then be washed by re-suspending them in a suitable fresh solution, centrifuging them to the bottom again, and then pouring away the solution.

In this way samples of pure viruses were first obtained, allowing chemical analyses to be performed to find out at last what the viruses are really made of. In every case the answer appeared to be largely two major types of chemical – proteins and nucleic acids[7,8] – which are also the two most basic types of chemical found in all forms of life. As the purification techniques improved, a number of other chemicals were shown to be lesser constituents of the viruses. The most prominent of these were materials known as "lipids", but a small amount of carbohydrate was also found, in addition to other very minor components.

The identification of the chemical composition of viruses is a great landmark in the history of virology. It ended the era of complete mystery and brought the viruses out into the open as biochemical entities accessible to detailed study. Before we can examine the fruits of these studies, however, we must take a look at the four basic types of chemical from which the viruses are made – the nucleic acids, proteins, lipids and carbohydrates. This will leave us well equipped to investigate how these chemicals are packaged together to form a virus, and how they allow the viruses to do such damaging and annoying things to us.

CHAPTER 2

Building-blocks[1] – of viruses, mice and men

One of Beijerinck's most important discoveries was that viruses can only multiply when they are "incorporated into the living protoplasm of the cell". We now know that the reason for this dependence on living cells is that it is the infected cell that actually *makes* new viruses. Indeed, the process of viral infection is a form of biological hijack, in which viruses invade suitable host cells and commandeer the complex molecular machinery of the cells to manufacture many more virus particles. So the viruses are really *products* of cell metabolism, albeit sometimes very damaging ones.

This intimate relationship between viral multiplication and normal cell metabolism will be explored more thoroughly later on, but in the meantime you should appreciate that it makes the basic molecular activities of the viruses very similar to the activities of their host cells. So it is impossible to understand how viruses work without also being aware of how normal cells work. To examine the nucleic acids, proteins, lipids and carbohydrates of the viruses then, we must consider the production and activities of these chemicals in the cells infected by the viruses. So let's take a look at these truly remarkable chemicals that are the building-blocks of both cellular and viral life.

First, a summary

The class of chemicals that scientists call nucleic acids includes the most celebrated molecule of them all – DNA (deoxyribonucleic acid). The magical acronym "DNA", and the fact that DNA molecules can coil together into a "double-helix", are familiar to millions of people who nevertheless know little about what DNA does or how it does it. In fact, the importance of the nucleic acids (including DNA) to all forms of life can be stated very simply: They are the "*genetic*" material of living things; which means that they store the information needed to construct and maintain all organisms, and which is passed on to subsequent generations. So the nucleic acids are the molecular agents of heredity – they ensure that all mice look like mice, that all men look like men and that all viruses look like viruses. But although they can produce endless generations of near-identical organisms, the nucleic acids also have the ability to accumulate the gradual and useful changes in their structure that we believe to be the basis of evolution.

Despite their central role in all forms of life, the nucleic acids cannot construct a living organism all on their own. Just as a computer program is useless without a computer, so the nucleic acids cannot function as the agents of heredity unless they interact with the other chemicals of life; and the most fundamental of these other chemicals are the proteins – the chemicals that actually do the work of cell and organism building. Proteins can be regarded as the biological servants that carry out the demands of their nucleic acid masters. Specific nucleic acids direct the production of specific protein molecules, which then perform two main tasks. Firstly, they act as the catalysts (called enzymes) that allow all of the chemical reactions of life to take place. Secondly, the proteins serve as structural units holding organisms together and giving them shape and the ability to move. Sometimes a single protein molecule can play both a structural and a catalytic role, and several other important tasks are also carried out by the proteins.

Finally, amongst the many chemicals that are produced, processed and also broken down by the catalytic proteins known as enzymes, are the lipids and carbohydrates – the remaining two components of the viruses. These chemicals perform many accessory functions in living things, in addition to their major role as energy storage molecules and structural units of the various boundary membranes and walls of all cells and some viruses.

To see how these four basic chemicals can carry out their seemingly daunting tasks we must look at them all in a bit more detail.

DNA and RNA – the nucleic acids

The nucleic acids are long chain-like molecules made by joining together many smaller molecules known as "nucleotides". DNA is made out of only four different nucleotides, usually known simply by their respective initials – A, T, G and C (see figure 2.1). A stretch of DNA can contain any number of these nucleotides, linked together in any order (see figure 2.2) so the term "DNA" refers not to one specific molecule, but to an infinite variety of molecules all based on the same general plan. The DNA molecules that carry the genetic information of living things usually contain many thousands or millions of nucleotides strung out along their lengths in different sequences. Despite the size and apparent complexity of these molecules it is actually very easy to describe precisely any particular piece of DNA. All we really need to know is the sequence in which the nucleotides are arranged, which can be noted very simply by writing down the appropriate initials in the correct order. Writing "ATGCGGATCA", for example, fully describes a specific piece of DNA ten nucleotides long.

So much for the basic structure of DNA; what about the celebrated ability of two DNA chains to twist around one another to form the double-helix? (see figure 2.3). The double-helix is now firmly established as the emblem of modern biology, and it forms because each of the four nucleotides can form very weak chemical bonds with *one particular* type of nucleotide on another DNA strand. So if two DNA strands are matching or "complementary", having all the appropriate nucleotides opposite one another, then the double-helix will be held together by the combined force of many of these weak bonds. Double-helical DNA is used as the genetic material of all cells (animal, plant or bacterial) and some viruses. Other viruses use single-stranded DNA, while some store their genetic information in double or single strands of the related nucleic acid "RNA". All you really need to know about the structure of RNA is that it is identical to that of DNA apart from two subtle differences (see box 2A for details).

Having been introduced to the nucleic acids, you now need to know a little about how they work. There are two main properties required of any chemical that is to serve as the carrier of genetic information. Firstly, it obviously must be able to *contain* information; and secondly there must be some way for this information to be easily copied, so that when cells multiply by dividing there

BOX 2A – DNA

The long chain-like DNA molecule is made by linking together any number and sequence of four nucleotides known as deoxyadenosine phosphate (A), deoxyguanosine phosphate (G), deoxycytidine phosphate (C) and thymidine phosphate (T) (see figure 2.1). Each nucleotide is itself made up of three smaller chemical units – a phosphate group, a sugar ring and a nitrogenous base. If the phosphate group is removed from a nucleotide then a "nucleoside" is formed. The four different nucleotides of DNA have different

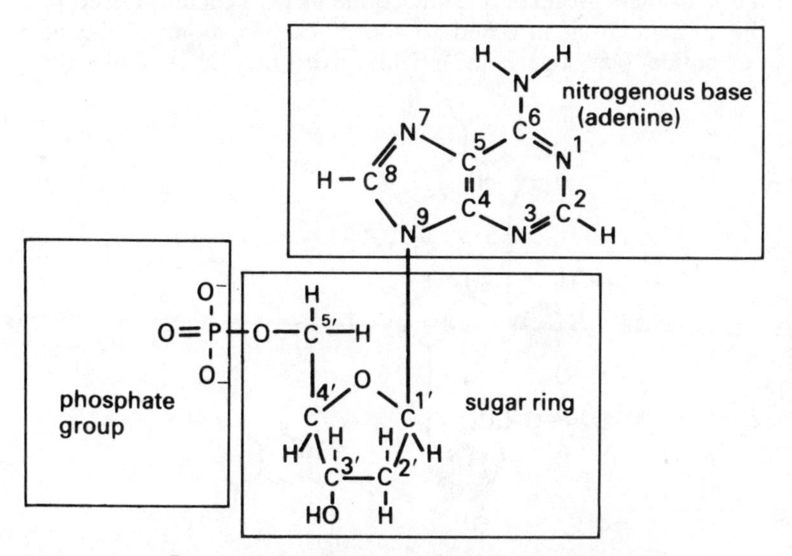

Deoxyadenosine phosphate (A) – a nucleotide

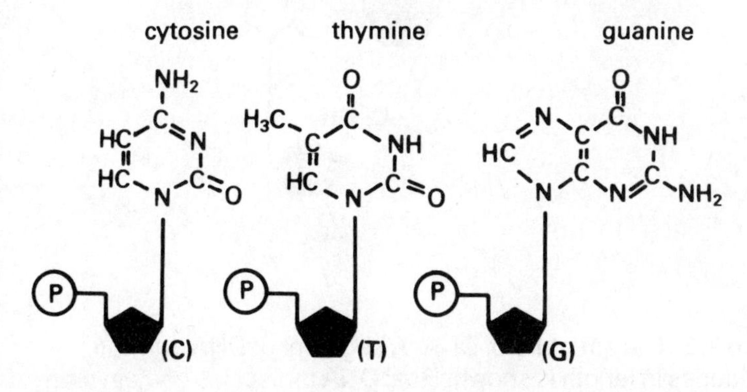

Figure 2.1 The four nucleotides of DNA

bases, but identical phosphate and sugar groups. So it is the *bases* that determine the unique characteristics of each nucleotide. The initials used to identify the nucleotides are strictly speaking the first initials of their component bases – adenine, guanine, cytosine and thymine.

The closely related nucleic acid called RNA uses a base called uracil (U) in place of thymine, and also employs a slightly different sugar ring. Apart from these two differences RNA is identical to DNA.

When nucleotides link up to form a stretch of DNA (or RNA), the phosphate group of one nucleotide forms a chemical bond with the oxygen atom attached to the 3′ carbon atom of the next nucleotide (see figure 2.2). This gives the DNA molecule a

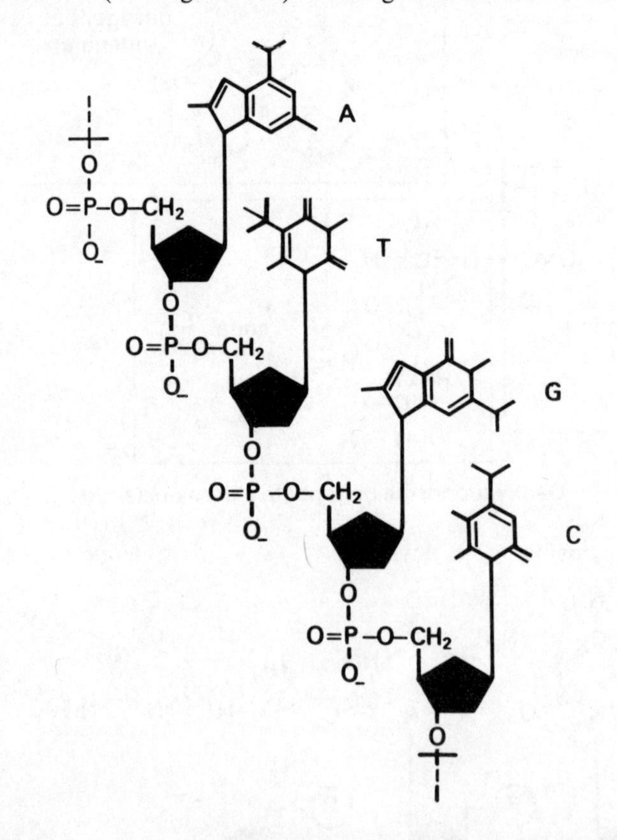

Figure 2.2 The structure of DNA. A segment of DNA only four nucleotides in length is shown. Real DNA molecules contain many thousands of nucleotides

constantly repeating "backbone" of sugar and phosphate groups, and a unique sequence of attached bases. The double-helix forms when two separate DNA strands become linked together by weak chemical bonds between specific bases. The rules of this "base-pairing" are as follows: adenine always pairs with thymine, while guanine pairs with cytosine. Two DNA molecules that can pair perfectly with one another, having the correct bases arranged in the correct order so that each base on one strand is opposite the base it can pair with on the other strand, are called "complementary" (see figure 2.3).

The base-pairing that holds the double-helix together also allows any specific double-helix to be readily replicated (see figure 2.4). This is because the information needed to construct any particular double-helix is in fact contained in either of the individual strands. If we know the nucleotide sequence of one

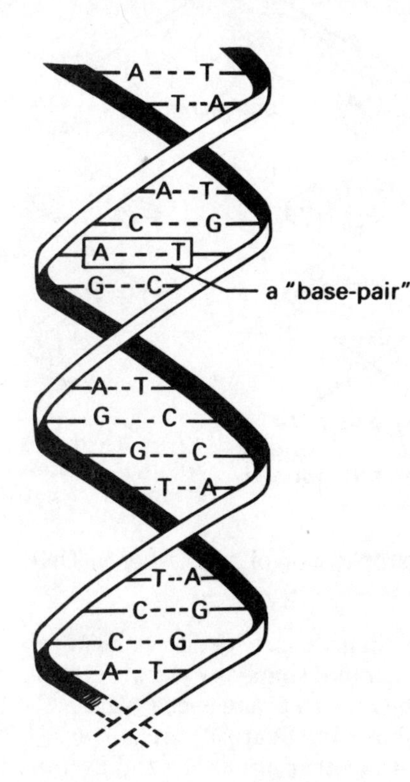

Figure 2.3 The DNA double-helix. Helical ribbons represent the sugar-phosphate backbones, initials represent the bases

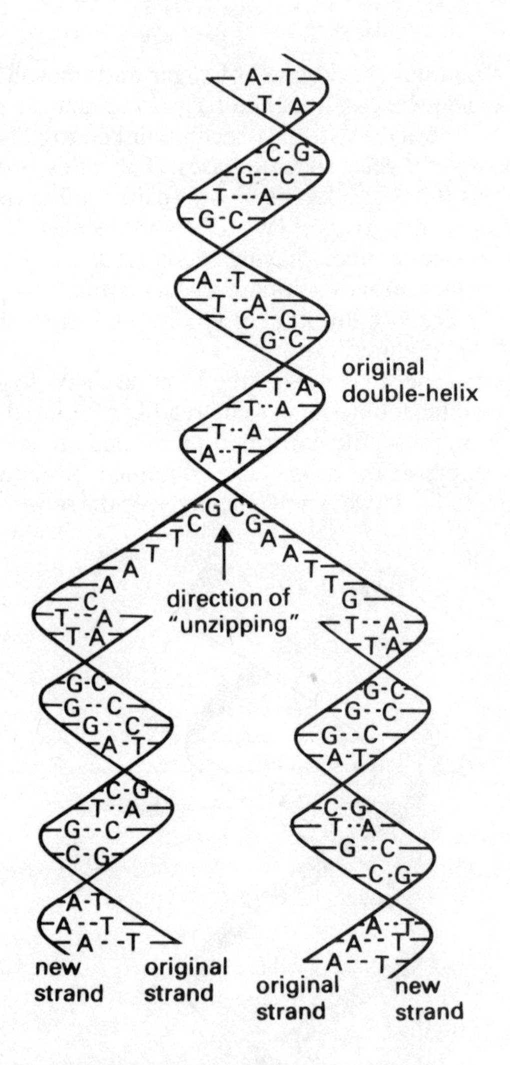

Figure 2.4 The replication of double-helical DNA

strand of DNA then we can readily work out the sequence of its complementary strand, simply by pairing each A with a T, each T with an A, each G with a C and each C with a G. The double-helix can unzip as shown in figure 2.4 relatively easily (because the bonds holding it together are weak) and as it unzips enzymes can link up the appropriate nucleotides into two new DNA strands complementary to the separating strands. Only correctly base-paired nucleotides will be incorporated, with the net result being the production of two identical copies of the original double-helix. It is this ability to undergo self-directed replication that allows DNA to carry genetic information from one generation (of cells or organisms) to the next.

The process of complementary strand formation can also be used to replicate RNA molecules, or to copy a single strand of DNA into a complementary strand of RNA, or vice-versa.

will be copies of the genetic information available to pass on to each of the two new cells. When James Watson and Francis Crick first discovered the double-helical structure of DNA in 1953, they realised at once that it offered a beautifully simple way for one double-helix to be copied into many more duplicate double-helices. If the original double-helix "unzips" as shown in figure 2.4, then the two separated strands can each serve as a template (a kind of "molecular mould") directing the production of a new matching strand. Remember that each nucleotide can only pair up with one other type of nucleotide, so the newly formed double-helices will be identical copies of the original.

Figure 2.4 illustrates how reproduction and inheritance work at a molecular level. The reproduction of all forms of life is ultimately dependent on "replication" of the genetic material by the process of matching or "complementary" strand formation outlined above.

We have seen, then, how the structure of the nucleic acids is well suited to one of the jobs they have to do – namely faithfully produce copies of themselves to pass on to future generations. But how do they manage to *contain the information* needed by these subsequent generations? Well we have also seen that the nucleic acids have an infinite capacity for structural variation, since the four different nucleotides can be arranged in any sequence along DNA or RNA molecules of any length. Anything that has the capacity for controlled variation can be used to store information, provided that some sort of *code* can be devised to relate the variations to the information stored. The words of this book, for example, are constructed from a set of 26 symbols which are themselves arranged into many thousands of words. But these marks on the page mean nothing unless you can decode them by knowing which marks correspond to which particular things and ideas I am trying to tell you about. The structure of the nucleic acids also contains information in the form of a code, and this "genetic code" is constructed from variations in nucleotide sequence. Most of the differences between you, me, a bacterium and a virus could be traced back to differences in the nucleotide sequence of our nucleic acids.

As has already been mentioned, the information stored within our nucleic acids is the precise instructions needed to make specific types of protein molecule. These proteins, particularly the enzymes, then go on to control the structure and activities of all living things. Before examining how the genetic code is *decoded* into proteins let's find out a bit more about the protein molecules themselves.

Proteins

The most important proteins of all are the enzymes. It is their job to make sure that the chemical reactions taking place in a cell are the ones required, and that these desired reactions occur in the right place, at the right speed and at the right time. Enzymes achieve their seemingly complex control over cell chemistry simply by acting as selective catalysts, greatly accelerating the rate of wanted reactions while giving no help at all to the undesirable ones. By definition, catalysts speed up chemical reactions while themselves remaining unchanged overall. So although an enzyme might be intimately involved in the chemistry of a reaction, it emerges from the reaction unchanged and ready to do the same job all over again. This allows a tiny amount of enzyme to wield a great influence over the chemical activity of a cell. It is our enzymes that actually make us into what we are, for without their help the chemistry of life would never proceed with the speed and control needed to keep us alive. The importance of the nucleic acids is that they carry the instructions needed to construct our enzymes (and other proteins), but it is the enzymes that actually do the work.

Just like nucleic acids, proteins are chain-like molecules made by linking together many smaller chemical units (see figure 2.5). The building-blocks of the proteins are called "amino acids", with twenty different types commonly found in proteins. All proteins begin life as a linear chain of linked amino acids (often several hundred), but most of them do not stay that way for long. Instead, the floppy protein chain very quickly folds up into what appears to be its lowest energy (most stable) conformation. Proteins that fold up in this way are known as "globular" proteins, due to the globule-like shape of their final folded form. Many working enzymes and other proteins are actually made up of several globular protein "subunits" held together by weak chemical interactions, and each subunit of such multi-subunit proteins is a separate folded protein chain.

The precise shape into which any protein chain becomes folded appears to be entirely determined by the type and sequence of its constituent amino acids (see box 2B). The folded structure of a protein in turn determines its chemical activities – what reactions it can catalyse and so on. The folding of an enzyme molecule, for example, usually creates holes and grooves on the surface of the enzyme into which only the chemicals involved in the reaction catalysed by the enzyme can fit. When attached to the enzyme in this way the reacting chemicals may be held in the precise relative orientation that allows the desired reaction to take place. Chemical groups on the enzyme itself can also participate in the reaction, although only transiently since overall the structure of the enzyme must remain unchanged. So the process of enzymic catalysis is dependent on the very precise interactions between an enzyme and the reacting chemicals involved.

BOX 2B – PROTEINS

Proteins are formed when amino acids become linked together as shown in figure 2.5, in what is known as a "condensation" reaction (since water is a by-product). The chemical bond linking the individual amino acids together is called a "peptide bond", and most proteins (also known as "polypeptides") contain a great many amino acids linked in this way. When only a few amino acids are linked together the product is called a "peptide", the only difference between a peptide and a true protein being one of size. The twenty amino acids are used to construct proteins are all identical apart from their "side-groups", represented by squares, triangles and so on in the figure. Once linked into a protein the amino acids form a long polypeptide "backbone" with the different side-groups strung out along its length.

It is the type and sequence of the side-groups that determines the shape a protein will fold into, and therefore what chemical activities the protein can perform. The different side-groups all have differing affinities both for the molecules in the surrounding

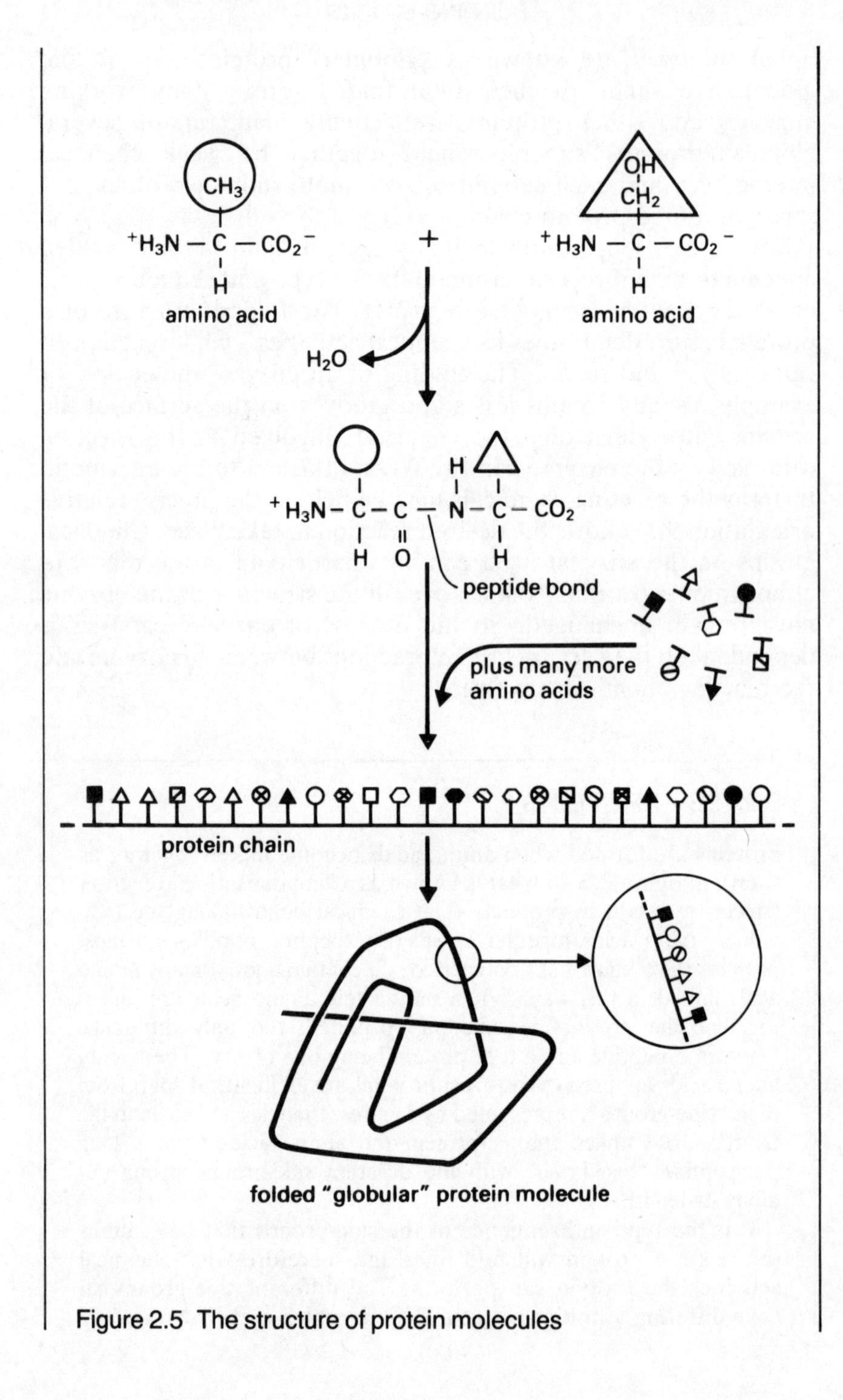

Figure 2.5 The structure of protein molecules

solution (mostly water) and also for one another. Some side-groups are strongly attracted to the watery environment around the protein, while others have no affinity for water at all and so tend to be pushed to the interior of the folding protein, away from the water. Similarly, some side-groups are attracted to one another while others are pushed apart by their incompatible chemical structures. All these competing forces seem to quickly resolve themselves to produce the particular folded arrangement that achieves the best possible compromise between the needs of the individual side-groups. This folded structure will probably be the lowest energy conformation of the protein, since it takes energy to push atoms into places where they interact incompatibly with one another. Some variation in the final three-dimensional structure may be allowed, but in general protein structure is highly specific.

The importance of protein folding is that it can bring together any combination of the available side-groups in an infinite variety of relative configurations. These clusters of side-groups are responsible for the chemical tasks performed by any protein. The differing structures of the side-groups allow enzymes, for example, to catalyse a bewildering range of different chemical reactions. There are side-groups that can act as acids and others that can act as alkalis, side-groups carrying a positive charge and others whose charge is negative, some that bind to water molecules and others that can hold on to “oily” hydrocarbons. Together they comprise a formidable chemical repertoire that makes all the complex chemical activities of living things possible.

Of course the chances of any amino acid sequence chosen at random resulting in a folded protein that acts as an efficient catalyst for any desired reaction are very small. But throughout the course of evolution countless numbers of alternative amino acid sequences must have been “tried”, with only the most useful being “selected” and slowly modified farther. The superbly efficient enzymes found in the organisms of today are presumed to be the select products of millennia of slow improvement – a process that no doubt often began with very coarse “enzymes” whose catalytic effects were only very slight.

Around the 1950s the amino acid sequence and precise three-dimensional structure of protein molecules was first beginning to be revealed. At about the same time the double-helical structure of DNA was first proposed and subsequently confirmed. In concert with these great breakthroughs a great deal of research was taking place into the relationship between the structure of the genetic material and the sequence and structure of the proteins that it coded for. How could a particular piece of genetic material (DNA in all cases except some viruses) direct the manufacture of a protein with a specific amino acid sequence? In other words, what is the nature of the genetic code?

The genetic code

The genetic code was broken by the combined efforts of several research groups, working throughout the 1950s and 1960s. The story of the many elegant experiments that eventually solved this central mystery of life could itself be the subject of a complete book. For our purposes, however, it is appropriate simply to summarise the results of these labours, results that allow us to trace in detail the flow of genetic information from the nucleic acid "library" through to working protein molecules.

We have seen that all the information needed to construct a three-dimensional protein molecule is contained within the linear sequence of the protein's amino acids. Once the correct amino acids are joined up in the proper sequence then basic thermodynamic forces do the rest, ensuring that the protein folds up into the precise conformation that allows it to do its job. So in order to contain the information needed to make proteins, all the nucleic acids really need to do is specify the sequence in which the amino acids of a newly forming protein will be arranged. Remembering that nucleic acids carry information in the form of variations in their nucleotide sequence, the problem of the genetic code can be reduced to a single simple question: *What is the relationship between the nucleotide sequence of a section of nucleic acid and the amino acid sequence of the protein it codes for*? The answer can be stated equally simply: *Each particular sequence of three nucleotides can specify the incorporation of one particular amino acid into a growing protein chain*. Both question and answer take only a few seconds to read, but they summarise a monumental amount of human effort and ingenuity. Let's take a look at how the whole process works.

For most organisms the sequence of events starts with the genetic information stored in the form of double-helical DNA (see figure 2.6), which usually contains the information needed to construct many different proteins. A section of DNA that contains the information needed to make one specific protein is generally called a "gene" and the entire gene library of an organism is its "genome". The human genome contains many thousands of genes which in turn encode thousands of proteins. Many viruses, on the other hand, contain less than 10 genes.

The copy of a gene that is stored in the form of DNA does not itself become involved in the enzymic nitty-gritty of protein manufacture. Instead, a working copy of the information is first made, which in all organisms takes the form of single-stranded RNA known as "messenger RNA" (mRNA). The production of the mRNA copy is achieved by separation of the twin strands of the double-helix, followed by the copying of one of them into mRNA by the same process of complementary strand formation as is used in DNA replication. For any particular gene the mRNA is a replica of just one strand of the double-helix, called the "plus" (+) strand, and the process of forming this mRNA replica is known as "transcription". This term continues the analogy with writing, with the genetic information being transcribed from the DNA master copy into mRNA containing the same information written in a different type of print (i.e. RNA instead of DNA).

All of the viruses that store their master copy of the genetic information in unconventional forms such as single-stranded DNA or double-stranded RNA, must eventually have it copied into single-stranded mRNA before it can be used to make protein. This is because the viruses rely entirely on the pre-existing machinery of the host cell to do the job of protein manufacture for them. In every cell on Earth the business of making proteins begins with mRNA.

The actual manufacture of proteins occurs when mRNA binds to complexes of protein and RNA in the cell cytoplasm known as "ribosomes". A ribosome moves along the mRNA molecule, and as it does so the sets of three nucleotides ("codons") that each code for an amino acid are exposed in turn at a special site on the ribosome. Each time a codon of three nucleotides is exposed at this site, the appropriate amino acid is brought to the ribosome and linked up into the growing protein chain (see box 2C for details). Once a ribosome has travelled the length of the mRNA molecule it will have produced the complete protein whose amino acid sequence is entirely and precisely determined by the sequence of

BOX 2C – HOW GENES MAKE PROTEINS

The genetic information contained within an organism's genome must be copied into mRNA before it can be used to make proteins. The mRNA binds to ribosomes in the cell cytoplasm and as the ribosomes travel along the mRNA the successive codons are exposed at a special site. Each time a codon is exposed at this site the appropriate amino acid is brought to the ribosome by another class of short RNA molecule known as "transfer RNA" (tRNA).

Transfer RNA is the "adapter" molecule that forms the crucial link between the structure of the nucleic acids and the structure of the proteins they code for. The tRNA molecules have two vital properties that allow them to perform this task. Firstly, each tRNA contains a region of three bases (the "anticodon") that is complementary to the mRNA codon that codes for one particular amino acid. This allows the tRNA to bind to the codon by base-pairing. Secondly, a separate region of each tRNA molecule carries the particular amino acid encoded by the codon it can bind to. So whenever an mRNA codon is at the correct site on the ribosome, an appropriate tRNA molecule pairs up with the codon and in doing so brings the correct amino acid to the ribosome. If the exposed mRNA codon is UGG, for example, the tRNA with ACC as its anticodon can bind to the UGG by base-pairing (remember the base 'U' is used in RNA in place of the 'T' used in DNA, so the A–T base-pair found in DNA is replaced by an A–U base-pair when RNA molecules are involved). This particular tRNA will carry the amino acid known as "tryptophan" (abbreviated to "Tryp" in figure 2.7), so the UGG codon in mRNA is responsible for tryptophan being brought to the ribosome.

As the ribosome moves along the mRNA, enzymes associated with the ribosome link up the newly arriving amino acids into the growing protein chain. Figure 2.6 summarises the whole process of gene expression, while figure 2.7 outlines which codons code for which particular amino acids. Most amino acids are encoded by more than one codon, and three of the codons do not actually code for amino acids at all. Instead of coding for amino acids, UAA, UAG and UGA serve as the signals telling the ribosome where it should stop the construction of a protein chain.

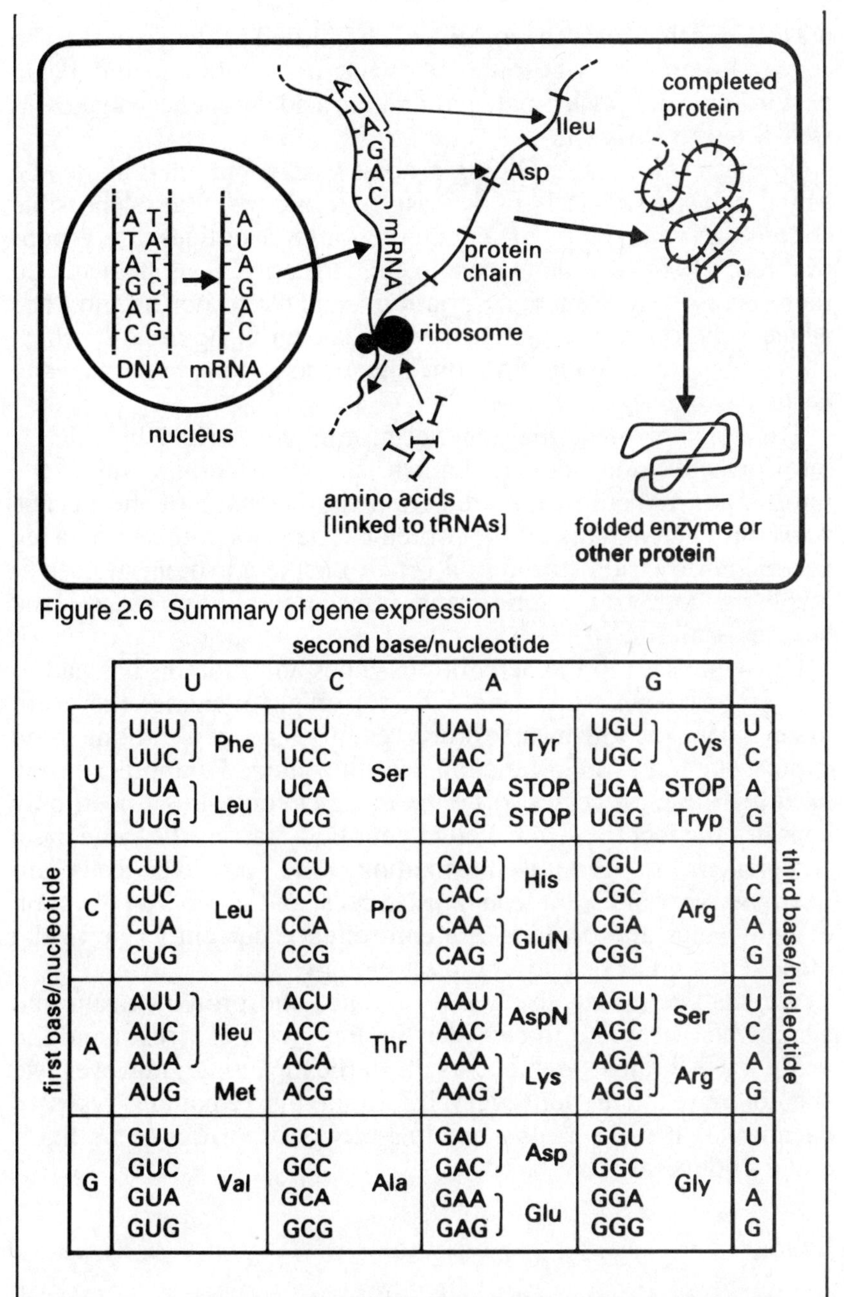

Figure 2.6 Summary of gene expression

first base/nucleotide	second base/nucleotide: U		C		A		G		third base/nucleotide
U	UUU	Phe	UCU	Ser	UAU	Tyr	UGU	Cys	U
	UUC		UCC		UAC		UGC		C
	UUA	Leu	UCA		UAA	STOP	UGA	STOP	A
	UUG		UCG		UAG	STOP	UGG	Tryp	G
C	CUU	Leu	CCU	Pro	CAU	His	CGU	Arg	U
	CUC		CCC		CAC		CGC		C
	CUA		CCA		CAA	GluN	CGA		A
	CUG		CCG		CAG		CGG		G
A	AUU	Ileu	ACU	Thr	AAU	AspN	AGU	Ser	U
	AUC		ACC		AAC		AGC		C
	AUA		ACA		AAA	Lys	AGA	Arg	A
	AUG	Met	ACG		AAG		AGG		G
G	GUU	Val	GCU	Ala	GAU	Asp	GGU	Gly	U
	GUC		GCC		GAC		GGC		C
	GUA		GCA		GAA	Glu	GGA		A
	GUG		GCG		GAG		GGG		G

Figure 2.7 The genetic code table. The amino acids specified by each codon are represented by their common abbreviations

nucleotides that form the mRNA. The finished protein will then be released from the ribosome, allowing it to fold up and begin performing the task that its amino acid sequence makes it well-fitted to carry out.

So that is how the decoding process works, but what about the actual code itself? It is fairly easy to work out that 64 possible codons can be made by arranging the four nucleotides into groups of three. These 64 codons have to code for only 20 amino acids, so there is more than enough variation available to do the job. The actual solution to the genetic code is shown in figure 2.7, which allows you to quickly find out the codons that code for any particular amino acid.

The conversion of the genetic information carried by mRNA into protein molecules is known as "translation", since the information is being translated from the language of the nucleic acids into the language of the proteins. The whole process of using genetic information stored in a gene to make a protein molecule, involving both transcription and translation, is known as gene "expression".

Our glimpse into the activities of genes and proteins has had to be very brief and superficial. As you might imagine, there are many ways in which the processes of gene replication, gene expression and protein function are controlled. To appreciate the *need* for these processes to be under strict control you need only consider the fact that every cell in your body carries the same basic complement of genetic information; and yet liver cells are obviously very different from muscle cells, and so on. The question of *how* genes and proteins are controlled is currently one of the most active areas of enquiry in all biology.

We shall return to the world of genes and proteins again and again when we come to consider the life-cycles of viruses and the ways in which they can make us ill. In the meantime, however, we should move on to look very briefly at the remaining types of chemical that serve as the building-blocks of viruses – the lipids and carbohydrates.

Lipids

The substances that biochemists call lipids are a diverse group of chemicals with one major property in common – they are not soluble in water. The term includes everything that a layman would call "fat". The lipids can be extracted from ground-up biological material by adding an organic solvent such as chloroform, ether or benzene. The watery remnants of the ground-up cells and the organic solvent will not properly mix, but if the separate layers are vigorously shaken together and then allowed to separate out again the lipids will tend to accumulate in the solvent. This can then be drawn off, allowing the lipids to be purified and analysed. The lipids obtained can be separated into several different categories according to their differing chemical structures, but taken as a whole they serve three major purposes in biology. Firstly, and most importantly as far as the viruses are concerned, they are the basic structural components of the thin membranes that surround all cells and some viruses. They also act as energy storage molecules (as many of us can testify by examining our waists!), being manufactured when an organism has an energy surplus and then broken down when energy is in short supply. Finally, the lipids play an important protective role in the skin of vertebrates, the exoskeleton of insects and the leaves of plants; making these exterior surfaces waterproof and insulated against the loss of heat.

The lipids that are found in viruses form a membrane, or "envelope", surrounding the virus. They can assemble themselves into membranes thanks to their general chemical structure which usually includes a long "fatty" tail, which is immiscible with water, and a short head-group made up of atoms that are strongly attracted to water (see figure 2.8). If lipid molecules with this type of structure come into contact with water, they tend to interact in ways that allow the needs of both parts of the molecules to be satisfied at the same time. To achieve this the "water-hating" tails must cluster together away from the water, while the "water-loving" heads need to be fully exposed to the water. One arrangement that meets both these requirements is shown in figure 2.8, the lower half of which depicts a cross-section through a spherical vesicle whose walls are a lipid "bilayer". By lining up back-to-back in this way the molecules keep all the tails surrounded by other tails, and so out of contact with the water either inside or outside the vesicle.

The membranes that enclose some viruses, as well as those found at the boundaries of cells, cell nuclei and various other intracellular bodies, are all composed of lipid bilayers. The forces causing these bilayers to assemble are very similar to those that force "water-hating" amino acid side-groups into the interior of a folding protein. In real biological membranes many different types of protein molecules can also be incorporated into the bilayer. These proteins sometimes seem to drift about in a sea of lipid, or they may be held at anchor in a particular place. They control the passage of material through the membrane and govern many

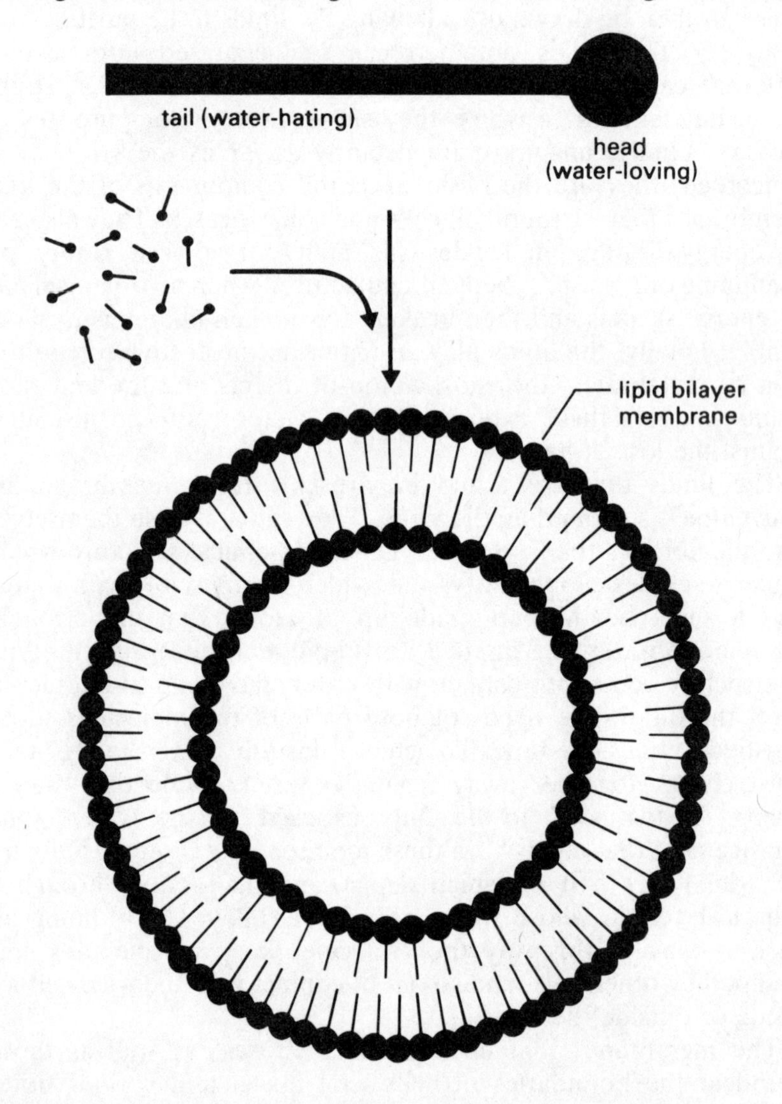

Figure 2.8 Generalised lipid structure and formation of a lipid bilayer membrane

aspects of the interactions between cells and their external environment. Although we will mainly be concerned with its function in viral membranes, the lipid bilayer is the structure that makes cellular life possible. It allows the contents of a cell to be strictly regulated and maintains each cell as an individual entity, distinct from its surroundings.

Carbohydrates

The carbohydrates are another diverse class of chemicals, the most familiar examples of which are the simple sugars such as glucose, sucrose (common "sugar") and fructose. Along with the lipids, carbohydrates serve as a major energy source for most organisms, and in addition to their role in energy supply and storage they are vital structural components of plant and bacterial cell walls. When performing structural or energy storage functions carbohydrates take the form of long chain-like molecules made out of many individual sugar molecules linked together.

Carbohydrates are only very minor components of viruses, although they may have a great impact on the overall behaviour of individual viruses. They are generally found joined on to certain viral protein molecules, forming hybrids of protein and carbohydrate known as "glycoproteins".

We need spend little time on the detailed chemical structure of the carbohydrates, but they are all composed of carbon, hydrogen and oxygen atoms and usually have the overall chemical formula of $(CH_2O)_n$. In figure 2.9 you can see the structure of the simple sugars mentioned previously, along with a small section of a chain-like energy storage carbohydrate called "amylopectin". Amylopectin is a major constituent of starch. Its structure demonstrates that (unlike most nucleic acids and proteins) carbohydrates commonly form *branched* chains.

With the carbohydrates our brief look at the chemical building-blocks of the viruses is complete. In this chapter the emphasis has been on how these building-blocks are formed and behave in the host cells that the viruses infect. It is time now to focus our attention onto the viruses themselves.

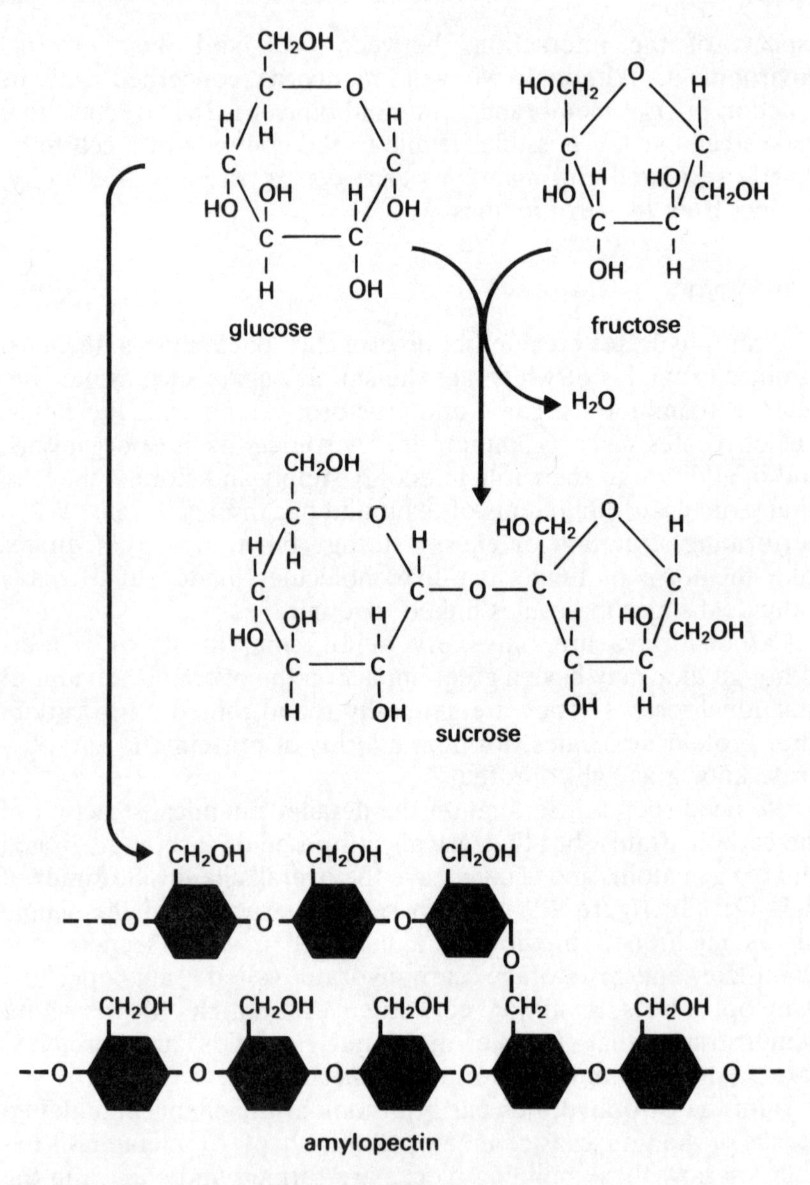

Figure 2.9 The structure of some carbohydrates

CHAPTER 3

Structure – the virus revealed

Arguments about what sort of infectious agents the viruses really were continued throughout the early decades of this century. There were those who agreed with Beijerinck that viruses were not cellular forms of life (indeed many people felt that they were not a form of life at all). And then there were those more cautious individuals who regarded the viruses as fairly normal micro-organisms – just very small ones. There was really only one way to resolve these disagreements – obviously the precise structure of a number of viruses would have to be worked out. But how was this to be achieved? It was impossible to see viruses through the microscope, and although details of their sizes and chemical content were very interesting they did nothing to solve the central dilemma of what do the viruses look like? The answers eventually came with the development of techniques that used electrons and X rays to "look" at viruses rather than rays of light; but a first clue was provided by an American biochemist called Wendell Stanley, working before even the chemical content of viruses had been fully elucidated.[1]

Stanley spent the early 1930s studying tobacco mosaic virus. At the time he thought that the virus was made entirely of protein, so it was natural for him to try to purify it using the techniques normally used to purify proteins. Pure crystals of proteins can often be obtained by treating impure protein solutions with salts such as ammonium or magnesium sulphate, and when Stanley added ammonium sulphate to fluid taken from plants infected with tobacco mosaic virus, small needle-like crystals of the virus slowly formed. He knew that the crystals were made out of virus particles because when they were collected and washed they could pass the disease on to the other plants.

Stanley's achievement in producing crystals of tobacco mosaic virus was yet another demonstration that viruses often behave more like chemical compounds than living organisms; and, more importantly for our present purposes, it showed that the tobacco mosaic virus behaved very much like a normal protein. When it later became known that the virus was not made entirely of protein, but contained nucleic acid as well, then its ability to crystallise like a protein gave a strong clue to its overall structure. It suggested that the protein might form an "outer coat" of the virus, with the nucleic acid somehow hidden away inside it. With such an arrangement only the viral protein would be exposed to the surrounding chemical environment, allowing the virus to behave much like a pure protein, just as Stanley had found.

This picture of the viruses as pieces of genetic material encased within protein coats was eventually confirmed using the powerful techniques of electron microscopy and X-ray diffraction. The electron microscope uses beams of electrons to do the job performed by rays of light in conventional microscopy. The electrons are scattered to differing extents by the various regions of the object being examined. The electrons that manage to pass straight through are then focused by magnetic fields (replacing the lenses of the light microscope) to produce an image of the object on a photographic plate. The regions of light and dark that make up this image reflect the differing degrees of electron scattering that take place as the electrons pass through the different parts of whatever is being examined. The superlative magnifying power of the electron microscope not only makes whole virus particles visible, but can also show up some of the structural details of the viral coat (see plates 3.1–3.4).

To look even closer, right down to the level of individual atoms, X-ray diffraction must be used. This involves firing X-rays at the object of interest, which must be in a crystalline or other highly ordered form. The interaction of the X-rays with the atoms of the object induces complex patterns in the emerging X-rays that can be interpreted by experts to reveal the relative locations of individual atoms. Unfortunately X-ray diffraction cannot produce photographs of viruses, genes, proteins and so on that could be understood by anyone, but drawings or computer-simulated pictures based on the X-ray data provide the next best thing. The most celebrated achievement of X-ray diffraction is Watson and Crick's initial discovery of the DNA double-helix.[2]

The combined techniques of electron microscopy and X-ray diffraction at last revealed many of the details of virus structure.[3] The views obtained confirmed that the viruses are certainly not small cells, but that they are indeed small pieces of genetic material (DNA or RNA) wrapped up inside a protein coat and sometimes further enclosed within an outer lipid membrane. These generalisations are confirmed by the detailed diagrams of virus structure given in figures 3.1–3.4. Tobacco mosaic virus (figure 3.1) has an extremely simple structure and no outer membrane. Just as Stanley's crystallisation experiments suggested, the protein forms an outer coat completely encasing the nucleic acid. Clearly, the coat is not one continuous layer of protein, but is instead composed of many identical globular protein "subunits". Even for a virus, tobacco mosaic virus is extremely simple, being built out of a single strand of RNA and just one type of protein subunit. Although other viruses can contain more complex forms of nucleic acid, may have an outer coat made up of a number of different types of protein (often with attached carbohydrate), and may be surrounded by a lipid membrane, they all adhere to the same overall plan of nucleic acid wrapped up inside a multi-subunit protein coat.

There is another general rule of virus design that sharply distinguishes them from cellular forms of life: viruses of any one type are always the same size and shape. Such homogeneity is not found in cells, particularly because of the way they multiply – slowly growing until they split into two separate smaller cells that then themselves begin to grow. The constant identical size and shape of properly formed viruses of any one type indicates that viral multiplication does not proceed by gradual growth followed by division. Instead, it suggests that they might be assembled from pre-formed components, much like a car or television set – a suggestion we shall see confirmed in the next chapter.

Some complexities

It is clear from figure 3.1 that the architecture of tobacco mosaic virus is based on the helix. The protein coats of most viruses tend to conform to one of two basic architectural plans, with the alternative to a helix being a roughly spherical shell of protein

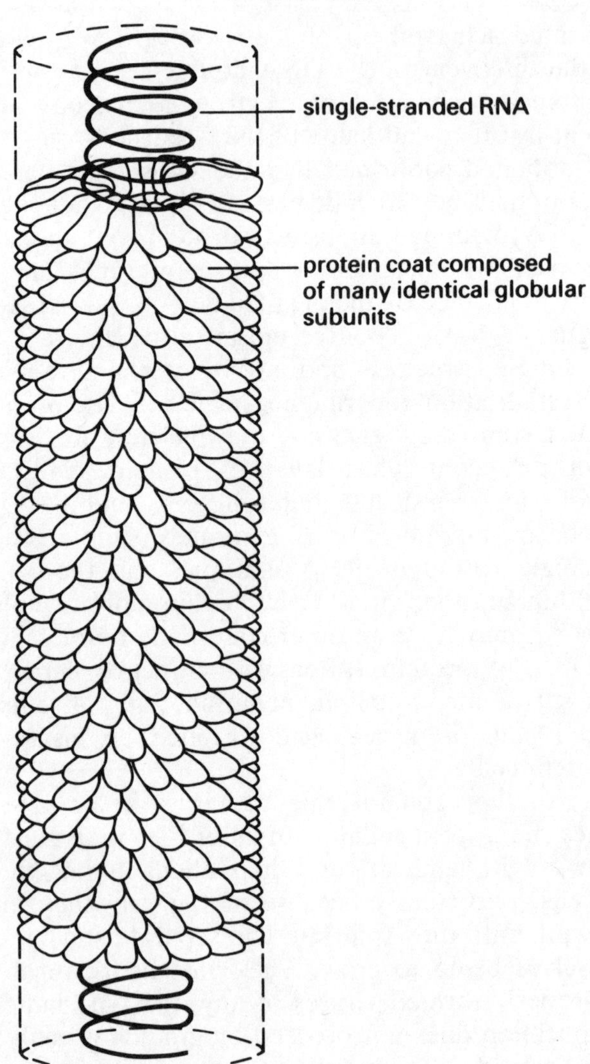

Figure 3.1 The structure of tobacco mosaic virus (the complete virus particle is about 6 times the length shown here)

subunits completely enclosing the central nucleic acid. In fact, the so-called "spherical" viruses are more accurately described as "icosahedral". (An icosahedron is a solid body with 20 triangular faces and 12 vertices. More complex structures that retain icosahedral symmetry can be built by dividing each face of the icosahedron into further roughly equilateral triangles.) The classic icosahedral viruses are the adenoviruses, which cause many of our colds and sore throats by infecting the cells lining the respiratory tract. The structure of adenoviruses (see figure 3.2 and plate 3.2) is also of interest to us because their outer coats are composed of several different types of protein subunit and, like many viruses, they have protein "fibres" or "spikes" protruding from the main body of the virus.

There are good reasons for viruses to be built in the form of icosahedra. The icosahedral shape allows the protein subunits to be tightly packed around a central space (containing the nucleic acid) in the most economical way possible.[4] In other words, it allows the viral nucleic acid to be completely enclosed by the minimum number of protein subunits.

Although adenoviruses are made out of several different types of subunit, they are still fairly simple viruses. To get an idea of what some of the most complex viruses look like we should consider the structure of influenza viruses, shown in figure 3.3. The influenza virus nucleic acid (single-stranded RNA) is first surrounded by protein subunits to form a flexible helical complex. In fact, there are eight such complexes within each virus, giving the virus what is known as a "segmented" genome, with the total genetic information being encoded by a number of separate RNA strands. The helical protein/RNA complexes are further enclosed

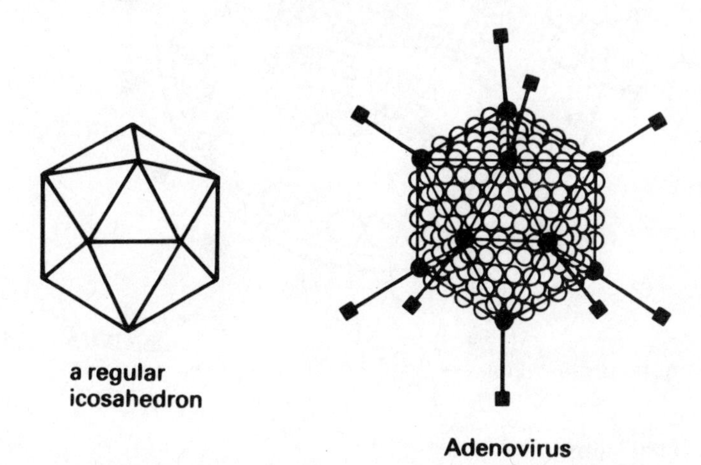

Figure 3.2 Icosahedral viral architecture

within an outer shell composed of a different type of protein subunit, and outside this shell comes the lipid bilayer that forms a membrane surrounding the entire virus. Embedded within the outer membrane there are glycoproteins that carry the carbohydrate groups of the virus. These glycoproteins form spikes that are involved in the interaction between the virus and the membranes of the cells it infects. Of course even such a "complex" virus is still much simpler than the simplest of living cells. It contains no cell cytoplasm, no protein-manufacturing machinery and is totally devoid of all the complex interacting metabolic processes (catalysed by enzymes) that make all cells work.

There is little point in considering the structure of a long list of viruses, since by now you should be well aware of the overall plan – all viruses are a piece of genetic material encased within one or

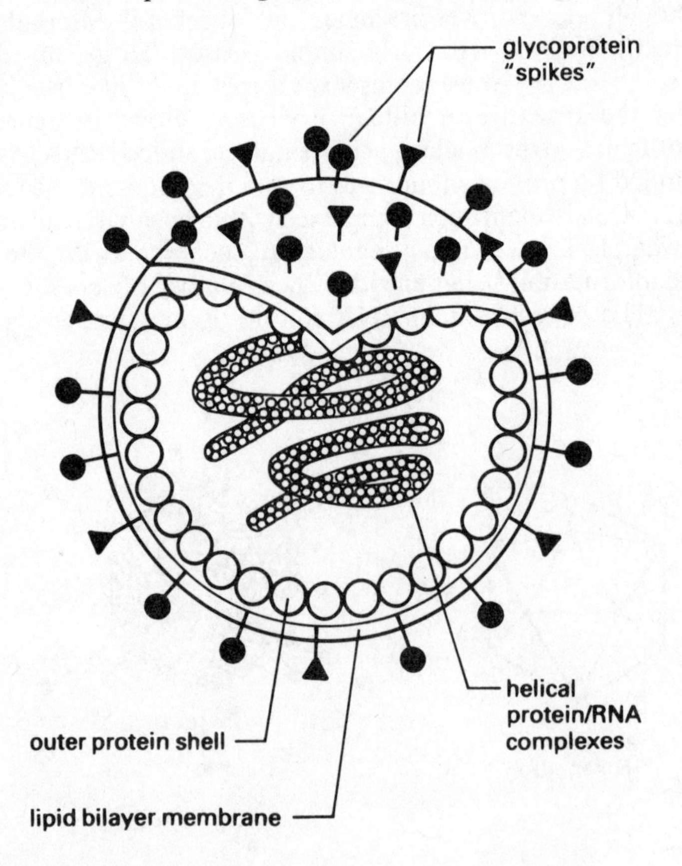

Figure 3.3 The structure of influenza virus

more layers of a protective outer coat. Although these outer layers can greatly influence the interaction of viruses with their host cells (indeed they can *determine* which types of cell become infected), the fact that they mainly serve a protective function can be proven by experiment. Purified RNA from tobacco mosaic virus, for example, can be used to set up a normal infection in the absence of any viral proteins, and despite the fact that in natural infections both the viral protein and nucleic acid enters infected cells.[5,6] Cells artificially infected with the purified RNA will eventually produce a crop of new viruses, made out of nucleic acid and protein as usual. So it seems that in the case of tobacco mosaic virus the protein has no real role to play other than ensuring that the genetic material is carried safely between, and can successfully penetrate into, the cytoplasm of the host cells. The proteins carried into cells by some other viruses *are* required for a successful infection to be set up. In these cases though, the proteins will have been made by expression of the viral genes that code for them during the previous cycle of infection, so the central importance of the genetic material still stands.

The unusual habits of the bacteriophages also confirm the belief that the nucleic acid is the really crucial part of a virus. Consider, for example, a rather complex class of bacteriophages known as the "T-even" bacteriophages, whose general structure is outlined in figure 3.4. These viruses are made up of a "head" region with icosahedral symmetry (and which contains the nucleic acid), and a helical "tail". The viral proteins also form a "collar" between the head and tail, and a "baseplate" at the bottom of the tail. The presence of a number of long protein fibres protruding from the tail gives these viruses an uncanny resemblance to the lunar module of the Apollo space flights, and the virus does in fact "land" on its host cell in a manner consistent with this analogy (see figure 3.4 and plate 3.4). After attachment of the fibres to the bacterial cell surface (remember, bacteriophages only infect bacteria) the baseplate is brought down and the DNA genetic material is "injected" into the cell. To allow this to occur the protein sheath of the tail contracts, driving the DNA down a central tube and into the cell cytoplasm. The entire protein coat is left outside on the cell surface and the DNA alone goes on to bring about the manufacture of many more viruses. So this time the normal viral life-cycle demonstrates that it is the genetic material that is the absolutely crucial part of the virus.

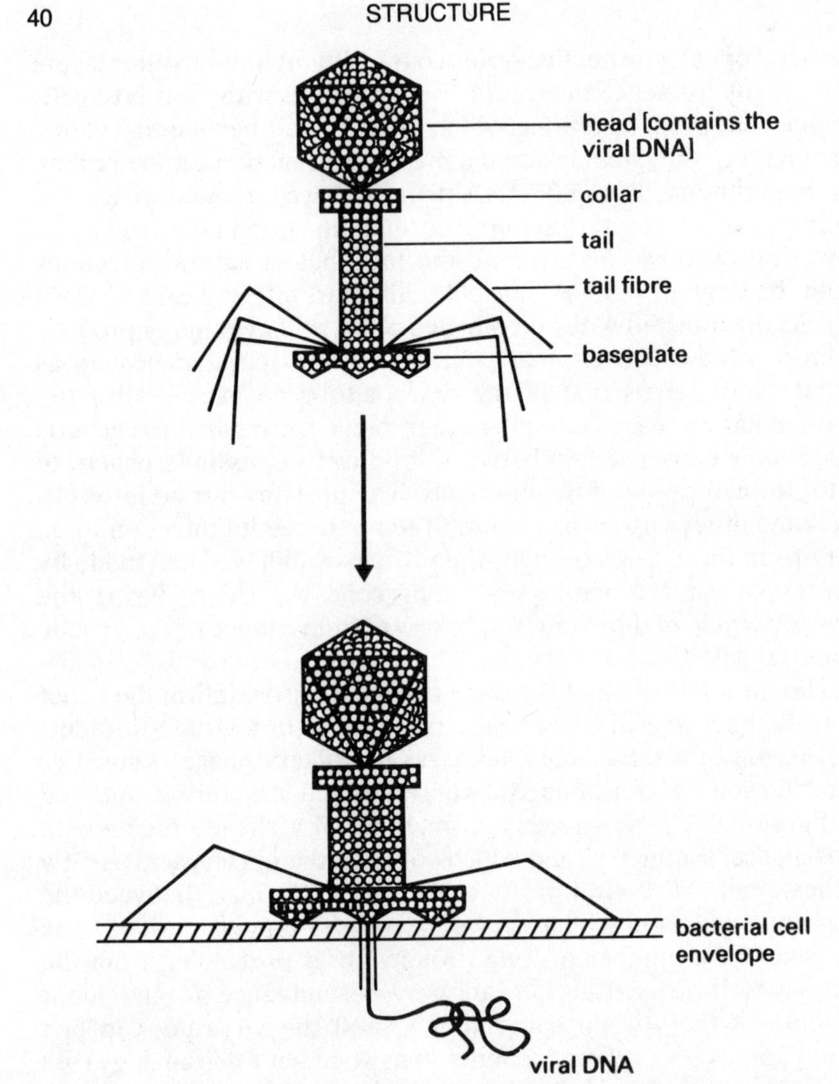

Figure 3.4 Generalised structure of the "T-even" bacteriophages and injection of the viral DNA into a bacterial cell

By looking at the structure of these four different types of virus you should have gained a good overall impression of viral architecture, and the scope for variation within the same general plan. In addition to the helical and icosahedral viruses there are a few whose protein coats do not fit either of these designs. These are called the "complex" viruses, a name that refers more to the complex symmetry of the protein coat than to any great complexities of the viruses overall. The division into helical, icosahedral and complex architectures is one very broad means of classifying the viruses. These three categories can be subdivided further by taking other physical features into account, such as the type of nucleic acid and the presence or absence of an envelope. The system of viral classification now in use is based mainly on these sorts of structural considerations, but the names used for individual viruses are less logically chosen. Some viruses are known simply by the name of the diseases they cause, giving us "influenza virus", "poliovirus", "tobacco mosaic virus" and so on. But other viruses have been named after their discoverers, the area of the world in which they were discovered, or even simply assigned a title from a letter and number code. The bacteriophages, for example, are known only by the codenames "T4", "T7", "c16" and so on. In figure 3.5 the main "groups" of viruses responsible for human disease are shown, along with some important members of each group. All of the viruses that are members of the same group, such as the herpesviruses, can be assumed to have a very similar structure, although the diseases they cause may be very different.

Looking down figure 3.5 tells us that the viruses are responsible for a wide spectrum of illnesses with consequences ranging from trivial irritation to rapid death; and the list only includes diseases for which the evidence of viral involvement is watertight and the role of the viruses straightforward. It ignores many other serious and often mysterious conditions in which the viruses might play some part (see chapters 10 and 13), and of course the viruses also afflict our pets, livestock and crop plants. The first step towards understanding how the viruses can cause such a diverse range of maladies is to explore exactly what happens when a virus enters a host cell and begins to multiply. How is it that a single tiny virus particle can bring about the production of many more identical viruses? How do viruses work?

Nucleic Acid	Shape	Envelope?	Group	Approx Size (nm)	Members of the group	Principal Diseases Caused
RNA	icosahedral	NO	PICORNAVIRUSES	24–35	Polioviruses Coxsackieviruses Echoviruses Hepatitis A virus Rhinoviruses	Poliomyelitis Aseptic meningitis Aseptic meningitis Hepatitis A Common cold
		NO	REOVIRUSES	60–80	Reoviruses Rotaviruses	? Gastro-enteritis
		YES	TOGAVIRUSES	40–70	Alphaviruses Flaviviruses Rubella virus	Encephalitides Yellow fever, dengue Rubella (German measles)
	helical [in spherical or bullet – shaped envelope]	YES	ORTHOMYXOVIRUSES	80–120	Influenza virus	Influenza
		YES	PARAMYXOVIRUSES	100–120	Parainfluenza viruses Mumps virus Measles virus Respiratory syncytial virus	Croup, other respiratory infections Mumps Measles Bronchiolitis
		YES	RHABDOVIRUSES	80–180	Rabies virus	Rabies
		YES	CORONAVIRUSES	80–130	Coronaviruses	Common cold
		YES	ARENAVIRUSES	110	Lymphocytic choriomeningitis virus Lassa fever virus	Lymphocytic choriomeningitis Lassa fever
DNA	icosahedral [herpes in spherical envelope]	NO	PAPOVAVIRUSES	50	Papilloma virus	Warts
		NO	ADENOVIRUSES	80	Adenoviruses	Respiratory infections
		YES	HERPESVIRUSES	120	Herpes simplex virus Varicella-zoster virus Cytomegalovirus Epstein-Barr virus	Genital herpes and cold sores Chickenpox, shingles Cytomegalic inclusion disease Glandular fever
	complex	NO	POXVIRUSES	230–300	Variola virus Vaccinia virus Orf virus	Smallpox Cowpox Orf
	not classified	NO		42	Hepatitis B virus	Hepatitis B

Figure 3.5 Classification of the more important viruses affecting man (*Adapted, with permission, from* A Short Textbook of Medical Microbiology *by D. C. Turk et al., Hodder and Stoughton, 1983*)

CHAPTER 4

Multiplication – life at the limit

Living things live in order to make more of themselves. In all biology this relentless single "purpose" of life is nowhere more evident than in the life-cycle of the viruses. With higher organisms such as ourselves it is possible to mask the central importance of our self-reproducing powers by pointing to such accessory activities as art, entertainment, academic endeavour and social interaction. Throughout our lives such activities often serve as alternative, more meaningful, "purposes" that keep us occupied and hopefully happy; but we are really only here because we can reproduce. With the viruses this "single-minded" driving force is about all there is. If they are to be considered a form of life (of which more later), then they represent life at its most basic, stripped of all accessories and built solely for the business of multiplication.

Viruses are little more than "mobile genes" surrounded by a protective coat. They carry a complement of genetic information that does nothing other than make proteins that directly assist in the reproduction and survival of the genes themselves. Of course ultimately, much the same description might well be applicable to even such proud organisms as ourselves; but the nihilistic futility of viral life is indisputably evident to all. Perhaps this is part of the reason why the viruses are often denied the outwardly grand title of "a form of life".

One of the "accessory activities" that has kept quite a few humans occupied throughout this century has been the attempt to solve the mysteries of the viral life-cycle. How can such tiny and simple agents multiply with a success that has resulted in the death of many millions of our own species? While much remains unanswered, we now have a good general grasp of the main events involved in viral existence. The variations and complexities that have been revealed make generalisation a daunting and hazardous task, but a task that must now be tackled.

Getting in

As long ago as Beijerinck's day it was realised that, in order to multiply, the viruses must somehow *get inside* the cells of a suitable host. But despite its long history, the problem of how exactly do viruses get inside animal cells remains a difficult and incomplete aspect of modern virology. It is obviously a question of great importance, since one of the best ways to counter the threat of viral infection might be to stop the viruses from getting inside our cells in the first place. A major problem in studying viral entry is that many encounters between viruses and cells do not result in any infection. In other words, many individual virus particles can bind to or even get inside cells, but all to no avail since they are soon eliminated by the cells' own "housekeeping" metabolism. So when viral entry is studied by electron microscopy or indirect biochemical means, it is often a considerable problem to separate the events that actually lead to successful infection from those associated with "non-productive" viral sorties into the cell. Despite these difficulties, however, some important general principles have been determined and several reasonable mechanisms for viral entry proposed.[1,2]

We have already seen the elegant way in which some bacteriophages can start an infection by injecting their genetic material into a bacterial cell. Infections of animal cells usually begin completely differently, with either the entire virus entering the cell, or at the very least the genetic material accompanied by some viral proteins getting in. Viral entry into plant cells, on the other hand, often takes place directly at sites where cell damage has been caused by the mouthparts of insects and other animals.

As far as entry into animal cells such as our own is concerned, the first and most basic principle is that the initial attachment of viruses to the cell surface is a highly specific process. It now seems clear that to start an infection viruses must first bind to particular protein molecules embedded in the host cell membrane. These "receptor" proteins are most likely to be glycoproteins that play some crucial role in the normal activities of the cell – they are certainly not there just to bind to passing viruses! Evolution has equipped the outer coat proteins of viruses with "binding sites" that can latch on to the cell receptors, forcing these normal cell proteins to act as unwitting doormen allowing viral entry into the cell. In the case of the enveloped viruses, in which the protein coat is surrounded by a lipid membrane, the protein spikes or fibres that protrude from the membrane carry the appropriate binding sites for attachment to suitable cells.

For successful infection to take place it seems likely that the initial link between virus and cell must be followed up by the formation of many more links between the binding sites and the receptors, and only then can the actual process of entry begin. Viruses are by no means efficient burglars of the cell, since only one out of every few thousand collisions between a virus and a suitable cell will actually result in tight binding. This initial step of becoming bound to the cell is apparently crucial to determining which viruses can infect which cells and therefore go on to cause disease. Any virus unable to bind to human cell membrane proteins, for example, will probably not be able to cause human disease; and even within our bodies the need for suitable receptors can restrict a viral infection to specific types of cell.

Once bound to the cell surface, there are three main ways for a virus to get inside, known respectively as "direct penetration", "endocytosis", and "membrane fusion" (see figure 4.1). A fourth and highly likely possibility is that some combination of these three basic entry mechanisms might be employed.

"Direct penetration" (see figure 4.1a) is a self-explanatory term. It describes the passage of a virus directly through the cell membrane, presumably accompanied by some loosening or "melting" of the membrane structure to let the virus through. Electron microscope photographs frequently show viruses apparently passing straight through cell membranes but how they might be able to do this remains a mystery.

"Endocytosis", on the other hand, is a well-documented process by which cells take up various external materials. It begins when the cell membrane folds inwards to form a pit engulfing the material destined for cell entry (a virus, for example – see figure 4.1b). The walls of this membranous pit then meet at the top, allowing a small vesicle containing the virus to break away from the membrane and move into the cell. This leaves the

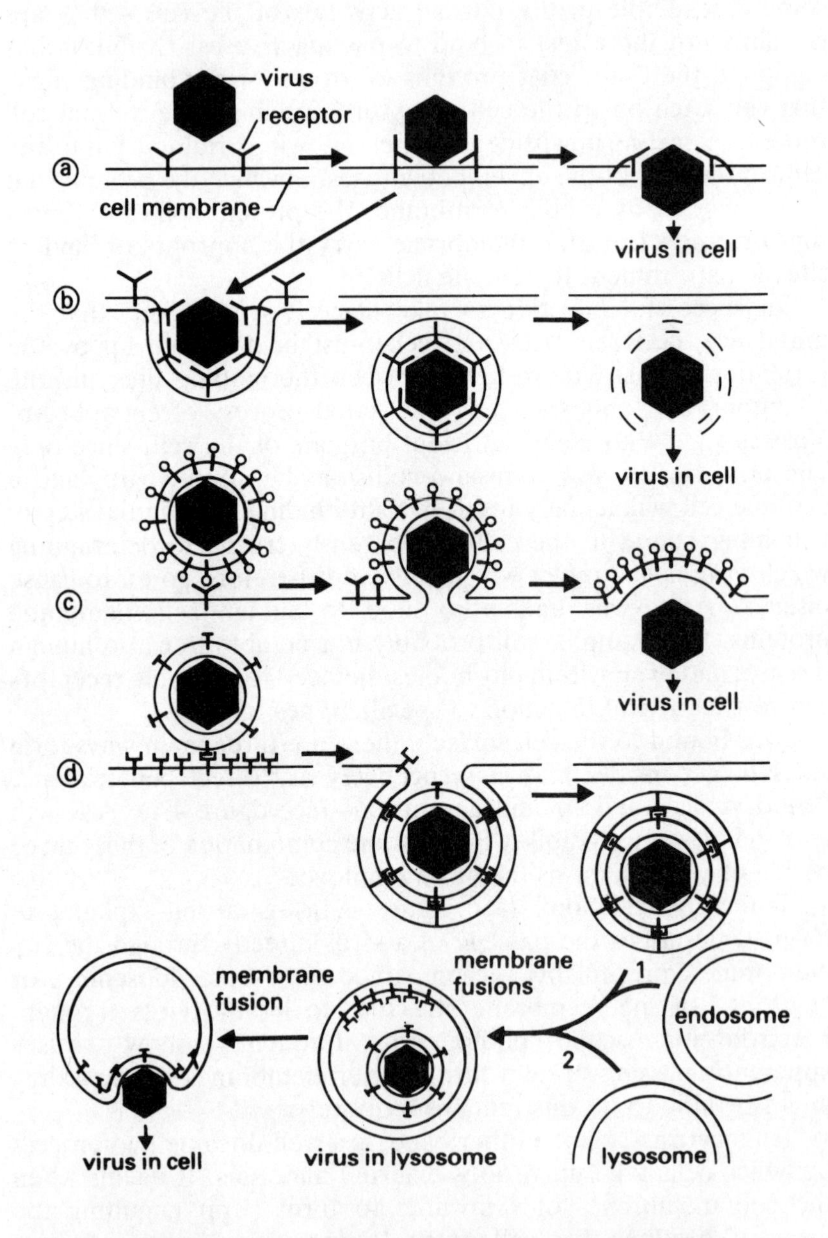

Figure 4.1 Mechanisms available for viral entry into animal cells. a) direct penetration through the cell membrane; b) Endocytosis; c) Fusion between viral and cell membranes; d) Proposed combined mechanism for the entry of influenza viruses

now enveloped virus inside the cell cytoplasm. It then only needs to escape from the vesicle, probably aided by the action of enzymes either carried by the virus or present in the cell, to leave it free to get the real process of multiplication under way. A multitude of studies using the electron microscope have confirmed that viruses certainly can enter cells by endocytosis (see plate 4.1), but as always the problem is to be sure that viruses entering in this way can actually cause infection once inside.

The third way for viruses to get inside cells, known as "membrane fusion", is restricted to those viruses that come supplied with their own lipid membrane. When two membranes touch they can fuse together into one continuous membrane (see figure 4.1c, releasing the main body of the virus particle into the cell. Again, much of the evidence in favour of fusion comes from electron microscope photographs that seem to have caught enveloped viruses in the act of membrane fusion (see plate 4.2).

The broad agreement among scientists, that direct penetration, endocytosis and membrane fusion are the three main mechanisms available for viral entry, contrasts with the considerable amount of disagreement over which particular mechanisms are used by specific viruses. Much of this debate might well prove to have been an unnecessary distraction, since most viruses might be able to enter in more than one way (although one particular route might lead more frequently to successful infection). Some of the most intensive investigations aimed at solving such problems have been carried out by Kai Simons and his research team at the European Molecular Biology Laboratory in Heidelberg. Simons' group have come up with a complex tale of endocytosis and membrane fusion to explain the entry of many enveloped viruses, including those that cause influenza.[3,4] So let's take a look at exactly what might be happening in the cells lining your nose and throat shortly before you miserably conclude that you have caught the flu.

First of all, according to Simons, the enveloped influenza virus enters the cell by endocytosis (you can follow the whole process through in figure 4.1d), having previously become bound to the receptors exposed on the cell surface. The endocytosis leaves the virus enclosed within a double-membraned structure, composed of the viral membrane itself surrounded by a membrane derived from the cell surface. The outer membrane of this double-membraned structure then fuses with the membrane of a cell body known as an "endosome", which then itself fuses with a similar body called a lysosome. So the two successive membrane fusions leave the influenza virus inside the lysosome and once again directly surrounded by only its own original membrane. Lysosomes are small membranous sacs containing powerful enzymes that can digest foreign material brought into the cell by endocytosis.

The virus can escape from the enzymes in the lysosome by a third membrane fusion, this time between the viral and lysosomal membranes. The inside of the lysosome is actually quite acidic and it is suspected that this acidic environment might alter the spike proteins of the virus, allowing them to promote the membrane fusion that releases the virus into the cytoplasm. At the end of the entire entry process, the virus will have reached the cytoplasm largely by taking a "free trip" on the machinery normally used by the cell to engulf and modify materials from its environment. This admirable economy on the part of the virus will emerge as a theme characterising many stages of all viral infections. If viruses could have a motto then an appropriate one would be "don't bother to do yourself what the cell can be made to do for you"!

Multiplication

Only after they have penetrated into the cytoplasm of a host cell do the viruses really get to work; and the most obvious effect of that "work" is the rapid production of many more virus particles, followed by their eventual release from the cell. This returns us to the problem, then, of how can the tiny viruses "hijack" a cell and convert it into a viral production line? Viruses have already been described as simply mobile genes. The amount of information carried by these genes is generally very small, being sufficient to produce at most 100 or so (and in many cases less than 10) different types of protein molecule. These *virally-coded* proteins are either the proteins of the viral coat, or else they are enzymes that catalyse specific steps in the cycle of viral multiplication. But for most viruses (even those encoding quite a few of their own enzymes), it is the enzymes of the infected cell that actually do most of the work involved in viral multiplication. In some cases an invading virus does not really *do* anything, its genetic material being expressed and replicated by cell enzymes simply because it is there. Cells, after all, are massive molecular factories devoted to the maintenance and replication of genes and the transcription and translation of these genes into proteins. Having gained entry into a cell, it is hardly surprising that viral genes can become incorporated into the manufacturing processes of these factories to eventually produce a new crop of viruses.

Many viruses, however, are not just passive beneficiaries of cellular activity. In addition to making viral proteins to help them multiply, many viruses encode proteins that actively interfere with cellular metabolism in ways that favour the needs of the virus. Such interference, for example, can either modify or perhaps shut down the manufacture of cellular proteins; and when viruses do interfere with cellular metabolism in this way it can profoundly influence the diseases they cause.

Leaving the subject of disease until later, it is fairly obvious that there are two main tasks to be accomplished during viral multiplication: firstly, the viral genes must be used to make the proteins they code for; and secondly, the viral genes must also be replicated so that copies will be available to incorporate into new virus particles (see figure 4.2).

Before either protein synthesis or gene replication can take place, however, the compact structure of the invading virus particle must be disrupted to make the all-important genetic material available to the enzymes that are required to act on it. For many viruses this "uncoating" (see figure 4.2a) involves a complete disintegration of the virus, releasing the viral nucleic acid into the cell; but for other viruses, particularly those whose coat proteins also act as enzymes, uncoating is more accurately described as a "loosening" process in which the purely protective proteins are removed, leaving a core of genetic material and viral enzymes. In some cases uncoating might be catalysed by enzymes either carried by the virus or present in the cell, while in other cases it may simply be the result of exposing a virus to the general chemical environment inside a cell. Regardless of how it happens, the end result of uncoating is to present the viral genetic material in a form in which it can be used both to make proteins and to be replicated.

With viral protein synthesis and gene replication we arrive at a point where the life-cycles of different viruses fragment into a bewildering complexity. Different viruses have different types of genetic material which can be transcribed and replicated in differing ways and can produce many different types of protein. So the variation involved in achieving the same overall aim is truly astounding. But don't panic! Fortunately the general principles are extremely simple.

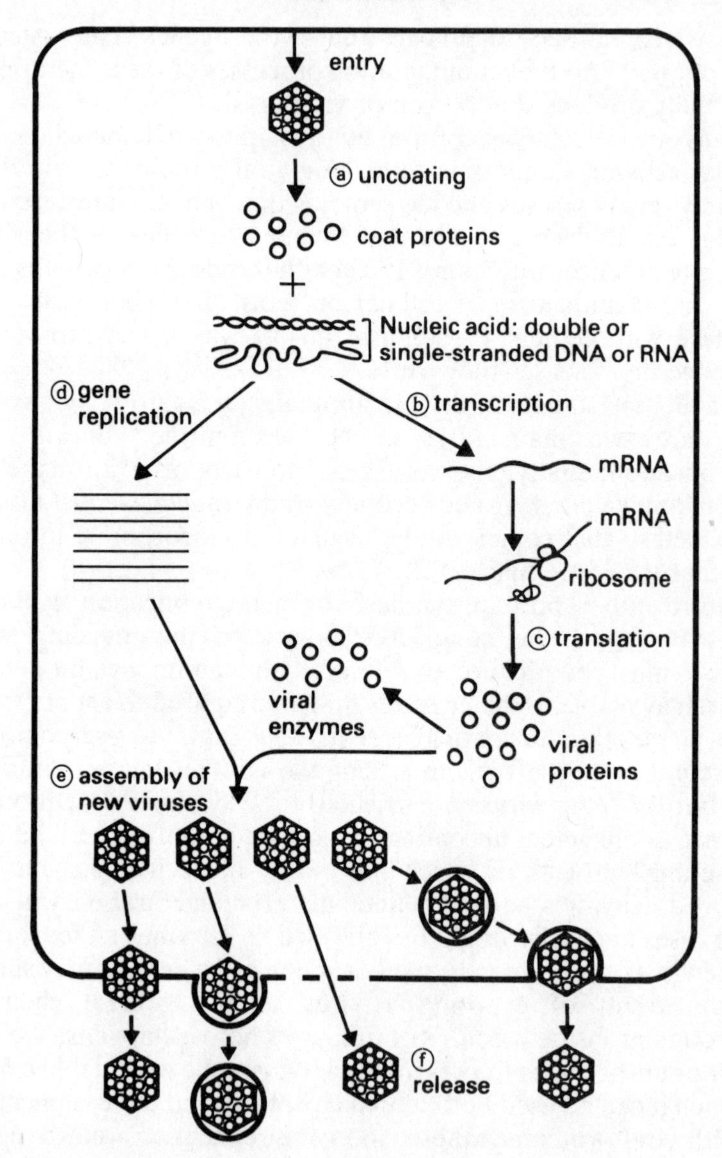

Figure 4.2 Summary of virus multiplication

As we saw in chapter 2, the genetic information carried by any piece of nucleic acid must be copied into a single strand of mRNA before it can be "decoded" to make protein. Some of the problems this presents to the viruses, and an outline to the solutions to these problems, are considered in box 4A. If you find such details confusing you can happily proceed armed with this one simple generalisation: In all viral gene expression the genetic material must be copied into mRNA, using the Watson and Crick rules of base (or nucleotide) pairing outlined in chapter 2; and the pre-existing ribosomes in the cell cytoplasm will then use this mRNA to produce the viral proteins (see figure 4.2b and c). Again, admire the economy of the viruses, exploiting the cell's own ribosomes and associated protein-making machinery to make the viral proteins.

Many of the ready-made enzymes that some viruses bring into a cell with them are needed to catalyse particular steps involved in making the viral mRNA. Some viruses *must* bring their own enzymes to do this job because it involves chemical events that do not normally take place within cells (the copying of a "minus" strand of viral RNA into "plus"-stranded mRNA for example). Once the viral mRNA has been produced then all of the protein molecules encoded in the viral genes can be quickly manufactured, including any required to assist in the second major event in viral multiplication – gene replication.

The process by which viral genes become replicated (see figure 4.2d), is similar in many ways to the production of the mRNA. In both cases the genetic information carried by a virus is being copied into the new nucleic acid molecules; but in gene replication an *identical replica* of the original genetic information is always the required end product. In chapter 2 we saw how identical replicas of the cell genome can be made, with double-helical DNA "unzipping" to allow the individual strands to serve as templates for the production of new complementary strands. Viral genomes made out of double-stranded DNA or RNA can replicate in much the same way. But with viruses using single strands of DNA or RNA as their genetic material, a complementary strand will first be made, which will in turn serve as the template to construct identical copies of the original genome. Just like mRNA production, the precise details of viral gene replication vary according to the type of virus involved; but in all cases the single simple overall mechanism is again the assembly of complementary nucleic acid strands according to the Watson and Crick rules of base-pairing. Throughout all the intermediary stages that might be involved, the viral genetic information is faithfully preserved in the nucleotide sequences of the newly forming nucleic acids.

BOX 4A – THE MANUFACTURE OF VIRAL mRNA

Viruses can be divided into six categories, depending on the type of genetic material they contain and the way in which the genetic material is expressed and replicated (see figure 4.3). For the

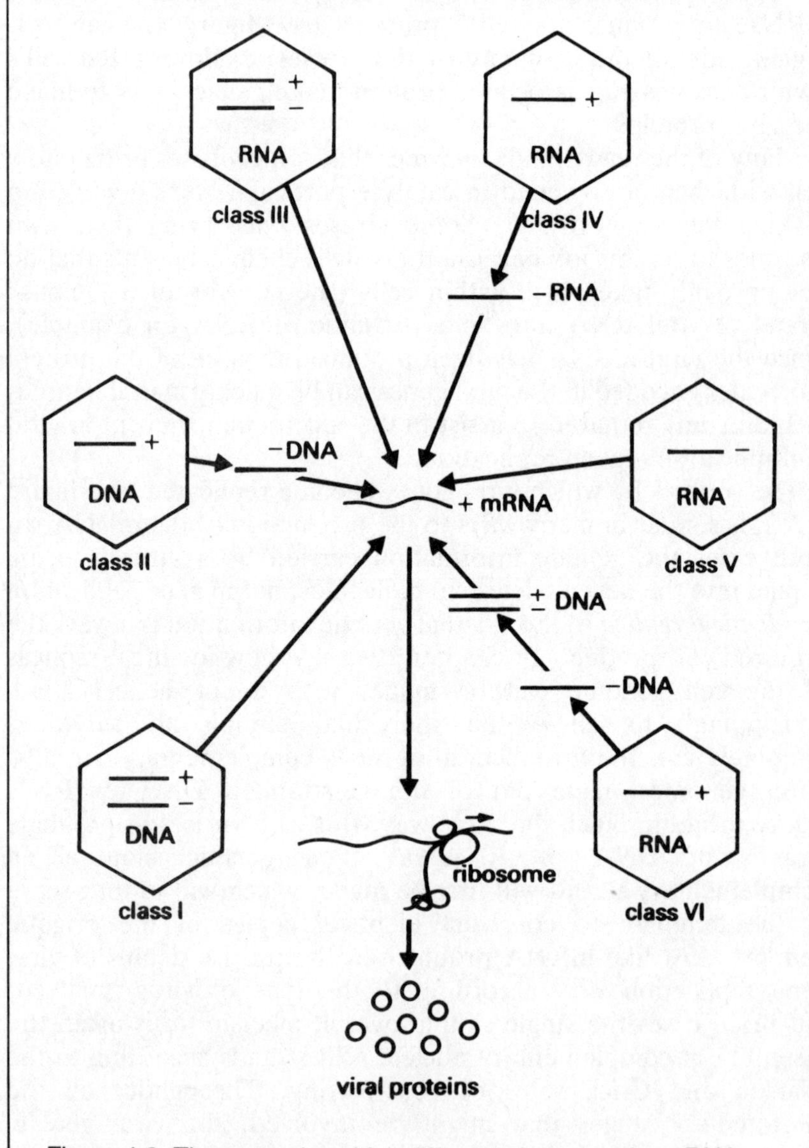

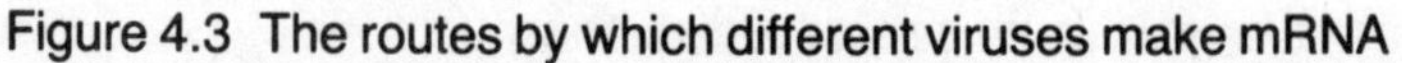
Figure 4.3 The routes by which different viruses make mRNA

genetic material to be used to make proteins its information must first be copied into mRNA, and for each category there is a different route by which the mRNA is manufactured. All of the required copying steps, however, are based on the production of complementary strands of either DNA or RNA made according to the Watson and Crick rules of base-pairing. These are the rules, remember, that allow the base sequence of one nucleic acid strand to determine the sequence of a complementary strand made using the original strand as a template (see chapter 2).

For the Class I viruses the process of mRNA production is very easy to understand. These viruses use the same form of genetic material as is found in all cells – double-stranded DNA. So to make mRNA the "minus" strand of the viral DNA is simply used as a template to make a complementary "plus" strand of mRNA. This is identical to what happens during the expression of normal genes belonging to the cell. (A "plus" strand of nucleic acid, remember, contains the actual sequence of nucleotides that can be decoded on the ribosome to make protein. A "minus" strand has a sequence complementary to this.)

In the Class II viruses the genetic material is single-stranded "plus" DNA, but shortly after infection a complementary "minus" DNA strand is made. mRNA can then be manufactured using this "minus" strand as a template in the usual fashion.

With Class III viruses the "minus" strand of the double-stranded RNA genetic material is used as a template to produce complementary mRNA.

The Class IV viruses would seem to have no problem in producing mRNA, since their genetic material is a single strand of "plus" RNA. You might reasonably expect that the genetic material could therefore be used itself as mRNA, but what actually happens is rather different. It seems that complementary "minus" strands of the viral genetic material are first made, which are then used as templates to produce the "plus" strands that actually serve as mRNA. Of course this system allows many more mRNA molecules to be made than the single "plus" RNA strand carried by the virus.

The single "minus" strands of RNA that serve as the genetic material for the Class V viruses can be used as templates to make complementary mRNA in one step; leaving us with only the Class VI viruses to deal with to complete our survey.

Class VI viruses have single strands of "plus" RNA as their genetic material, just like the Class IV viruses, but the similarities end there. Class VI consists of important and unusual viruses known as “retroviruses”. They display a complicated life-cycle in which their RNA is first copied into double-stranded DNA, followed by the incorporation or “integration” of this DNA into the host cell’s own genetic material. Consideration of the retroviruses is left until the next chapter, in which the whole topic of such integration is discussed.

What I am doing in this chapter is summarising the broad principles of processes that exhibit wide variations in their individual details. One additional possibility for variation is the *site* within the host cell at which the various stages of viral multiplication take place. Replication and transcription can occur either in the cell cytoplasm or the nucleus, depending on the type of virus involved. The cell’s own DNA, of course, is found within the nucleus, and it is inside the nucleus that normal cellular genes are replicated and transcribed. Similarly, most of the viruses that use DNA as their genetic material also undergo gene replication and transcription within the nucleus. One very good reason for this is that these processes often require cellular enzymes that normally replicate and transcribe cellular DNA, and which are therefore to be found in the nucleus.

The replication and transcription of RNA viruses, on the other hand, often takes place in the cytoplasm. Since the cell’s own genetic material is not RNA, replication and transcription of the RNA viruses often involves chemical steps that do not occur in normal cells. This means that specific viral enzymes must be used to catalyse these steps, enzymes which will either be brought into the cell with the virus or else manufactured shortly after infection.

The business of making viral proteins (translation) is thankfully less variable. In all cases it is the cell’s own protein-making machinery (ribosomes, transfer RNA, etc.) that converts the information carried by viral mRNA into protein, and all this machinery is found in the cell cytoplasm. So viral proteins are always produced in the cytoplasm.

Wherever the various events involved in viral multiplication take place,the raw materials required are all taken from the host cell. Thus the nucleotides to make viral DNA or RNA, the amino acids to make proteins, and any lipids or carbohydrates that are later incorporated during virus assembly, are all stolen from the cell’s own supplies. Viruses have been described as the worst possible "house guests" – arriving uninvited, raiding the host's larder and often leaving in their wake a trail of ruin or disarray!

Before we examine how all the newly manufactured viral genes and proteins manage to aggregate into new viruses and make good their escape from the infected cell, it will be worthwhile to take a quick look at exactly what sort of genetic information is carried by a few individual viruses, exactly what sort of proteins the viruses actually make. This will not only make the role of the viral proteins clearer, but should also give you a better impression of the quantity and range of genetic information that different viruses need in order to multiply.

Parvoviruses – a lesson in economy

Parvoviruses are probably the simplest true viruses to infect humans and other animals. They have been known for a long time to cause abnormal growth and development in rodents, but have only recently begun to be linked with human disease. They can apparently destroy human red blood cells, cause a childhood rash known as "erythema infectiosum", and may even have a role to play in rheumatoid arthritis.[5,6] The genetic material of a parvovirus is a tiny piece of single-stranded DNA, protected by just three different coat proteins. The DNA of one particular parvovirus (which infects mice) is known to be only 5081 nucleotides long and contains the information needed to make just four or five viral proteins, three of which form the viral coat.[7,8] The job of the other one or two proteins is not known, but even if they catalyse a couple of steps in the viral life-cycle it is clear that the virus is almost entirely dependent on the enzymes of the host cell to transcribe, translate and replicate the viral genome – all the tasks required for multiplication. The parvoviruses represent parasitism close to its limits, since they are little more than a piece of DNA encoding proteins that can surround and protect that DNA. They survive and multiply thanks to an ability to become passively integrated into the metabolism of the host cell, sometimes causing considerable damage in the process.

Influenza viruses – a bit more complex

The genetic material of the influenza viruses only codes for about

10 or 11 proteins, not many more than the parvoviruses, but some of these viral proteins catalyse absolutely crucial steps in viral multiplication.[9] Some of them actually have to be taken pre-formed into a cell along with the viral genes for a successful infection to be set up. As we have already seen, the influenza virus genetic material is split up into eight distinct segments of single-stranded RNA, each of which codes for at least one protein. Three of the viral proteins work together to make the viral mRNA, and these are the proteins that must be carried into the cell as part of the virus. Two of the other proteins form the glycoprotein spikes that stick out from the viral envelope, one forms the protein shell of the virus, while another is used as the protein that actually surrounds the individual segments of viral RNA (see figure 3.3 p. 38). The three or four remaining proteins encoded by the viral genes are not found in completed virus particles, so they presumably act as enzymes catalysing some stages of viral multiplication. Exactly which stages remains unknown.

Herpesviruses and poxviruses – many genes and proteins

The most complex viruses contain a large number of genes that in turn produce many different proteins. The herpesviruses for example – which cause many human illnesses such as chickenpox, shingles, genital “herpes” and cold sores – produce at least 50 of their own proteins from a complex genome of double-stranded DNA.[10] The poxviruses, responsible for smallpox for example, are probably the largest and most complex viruses of all. Again, they have a double-stranded DNA genome, this time encoding over 100 different types of protein.[11] Many of the proteins produced by these complex viruses are “structural”, being used to form the outer protein coat, but a large number are enzymes that catalyse many important steps in the expression and replication of the viral genes.

The ability to code for many of their own enzymes makes these viruses more *versatile* than the simpler viruses such as the parvoviruses, but it also makes them more *vulnerable* to our attempts to control them. The parvoviruses, for example, can only multiply in cells that are actively dividing. This is because the viruses are wholly dependent on the DNA-replicating enzymes of the host cell to replicate the viral genetic material, and large amounts of these enzymes are not found in non-dividing cells. Poxviruses, on the other hand, can not only multiply in non-dividing cells, but unlike most DNA viruses they actually multiply in the cell cytoplasm rather than the nucleus. So the ability to produce many of their own enzymes has freed them from any dependence on the enzymes found within the nucleus. The vulnerability of the more complex viruses arises because their uniquely viral enzymes provide targets for scientists trying to develop anti-viral drugs. A drug that interferes with a specific viral enzyme might be able to stop multiplication of the virus, while leaving the enzymes needed by the cell unaffected. Such an anti-enzyme drug attack is much more difficult against the simpler viruses, since they multiply by exploiting enzymes that are also essential for the growth and survival of the healthy cells of the infected organism.

So the genetic information carried by different viruses spans a wide range, making anything from just a few to over 100 viral proteins. Even the most complex viruses, though, are still dependent on the host cell to perform much of the chemistry involved in viral multiplication, particularly the complex business of linking up amino acids into viral protein molecules. There is no such thing as a free-living virus capable of surviving on a supply of simple nutrients. Outside of living cells the viruses are inert, helpless and harmless.

Assembly

In the hours following the entry of a virus into a cell, large quantities of newly manufactured viral proteins and nucleic acids will begin to accumulate. The assembly of these components into new viruses has already been likened to the production-line manufacture of cars or television sets. But industrial production lines need people, or at least robots, to fit the correct parts together and tighten the nuts and bolts. Many viral "production lines" have taken automation one step further, with the whole process of virus assembly proceeding spontaneously once the basic protein and nucleic acid components have been formed.

The fact that at least some viruses can assemble spontaneously is probably best demonstrated by studies on that familiar work-horse of virology – the tobacco mosaic virus. Tobacco mosaic virus, remember, consists of many identical protein subunits stacked into a helical coat around a central strand of RNA. If the purified virus is treated with appropriate salt solutions of carefully controlled acidity, then the entire virus structure falls apart, leaving a mixture of separated subunits and naked RNA. This mixture resembles the pools of viral components that build up in a plant cell during tobacco mosaic virus infection. By slowly changing the salt concentration and acidity the protein subunits can be made to clump together, first of all into single discs, and then eventually into completely reformed helical coats surrounding the viral RNA. So the virus can assemble completely spontaneously, provided the surrounding chemical environment is appropriate. Proof that such re-assembled viruses are infectious has been obtained by administering them to previously healthy tobacco plants.[12]

So, at least in principle, the assembly of viruses from their separate protein and nucleic acid components could simply be the result of spontaneous aggregation, presumably driven by the increased stability (lower energy) of the completed virus. This process would be similar in many ways to the spontaneous folding of protein chains into their final three-dimensional structure. In both cases basic thermodynamic forces allow a seemingly complex structure to be formed automatically from simple raw materials.

Many viruses other than tobacco mosaic virus are believed to assemble spontaneously, but studies on some viruses (such as bacteriophages and poliovirus) have shown that things are not always so simple. With poliovirus (and many other viruses), the proteins that eventually form the viral coat are first produced as large precursor proteins. At specific stages throughout virus assembly these precursors are chopped up by enzymes to yield the coat proteins of the finished virus. Other viruses produce proteins that are essential for virus assembly, but which do not themselves appear in the finished viruses. Such "processing" proteins must be analogous to the workers or robots on an industrial production line, perhaps manoeuvring some viral components into their correct respective orientations and holding them in place until other parts of the virus arrive.

Getting out

At the end of the viral production line, then, we will find completed viruses composed of nucleic acid surrounded by one or more layers of protective protein. Many of the assembly steps leading up to this stage will have occurred spontaneously, while some might have required the intervention of enzymes and other proteins produced by the virus or found within the cell. Of course some viruses have still to collect a lipid membrane before they leave the cell, and the membrane is often added as part of the final stage of a multiplicative viral infection – the release of newly made viruses from the infected cell.

There are at least four routes available for viruses to escape from cells; an escape that can be dramatic and destructive, or stealthy and certainly less damaging in the short term (see figure 4.2f). The most destructive path involves the death and disintegration of the host cell as a result of the viral presence within that cell. Viruses can interfere with cellular metabolism in various ways (see chapter 8), some of which can cripple the cell's own vital processes and lead to rupture of the membrane and cell death (a process known as cell "lysis"). The disintegration, or lysis, of the infected cell obviously allows the newly formed viruses to escape and carry the infection on to other healthy cells. In addition to being directly due to the effect of a viral infection within a cell, cell lysis can also arise when the body's defence systems attack an infected cell in an effort to eliminate the virus (see chapter 7). Many of the "naked" viruses (those not surrounded by a membrane) are released during cell lysis, but at least two alternative escape routes for such viruses exist. Some can apparently be extruded through the membrane in a reversal of the direct penetration entry path, while others (such as poliovirus) are released by a form of "reverse endocytosis". In reverse endocytosis the virus is incorporated into a small lipid vesicle within the cell, which then fuses with the cell membrane to dump the virus outside the cell.

Enveloped viruses often acquire their outer membranes as they escape from the cell by "budding" outwards from beneath the host cell membrane. This budding is really a reversal of the membrane fusion entry pathway and electron microscopy has provided impressively clear pictures of many viruses budding out and acquiring their membranes in this way (see plate 4.3). But the viruses are nothing if not versatile and variable, and so it should come as no surprise for you to learn that some viruses get their membrane by budding from the nuclear membrane (i.e. the one surrounding the nucleus) rather than from the outer membrane of the cell. The membranes of the enveloped viruses are not viral products, but are obviously stolen from the cells they infect. The proteins embedded within these membranes, however, *are* viral in origin. When these membrane proteins are first made they are passed into the cell membrane. On their way they can have carbohydrate groups attached to them by cell enzymes, producing the characteristic glycoproteins found protruding from the membrane of enveloped viruses. Once embedded in the membrane these viral proteins clump together and bind to the newly made viral protein and nucleic acid cores as they assemble in the cytoplasm below the membrane. The virus then buds out from the cell to adopt its final enveloped form. Obviously release by budding is much less damaging to the host cell than release via cell lysis. The cell membrane remains intact, allowing infected cells to survive for long periods of time while large numbers of viruses can be produced within them and released.

These then are the basics of viral multiplication. If you find any of the details confusing then remember the supreme simplicity of what is going on overall – *viruses simply enter cells; have their genes decoded into proteins and replicated to produce copies for incorporation into new viruses; and then the new viruses form by aggregation of the newly made genes and proteins and escape from the cell to start the whole process all over again.*

Alas, having hopefully impressed you with the simplicity of viral multiplication, I must now reveal to you another major variation on the theme of the viral life-cycle. It is a variation that allows an invading virus to truly take up residence within an infected cell, with the viral genes becoming an integral part of the host cell's own genetic material.

CHAPTER 5

Integration – the virus "lies low"

Trying to summarise the activities of the viruses is a bit like tracing the flow of rainwater off a mountainside – you have to highlight the role of the few streams into which most of the water flows, while remembering that small amounts of water can trickle and seep downwards by a great variety of lesser paths. The mainstream multiplicative pathway outlined in the last chapter is certainly not the only course that a viral infection can follow. In 1949 André Lwoff of the Pasteur Institute in Paris began a series of experiments that eventually revealed a second major path available to at least some viruses. It is a path which leads to the viral genetic information becoming incorporated into the genetic material of the cell, a process known as "integration".

Lysogeny

Lwoff was trying to solve a mystery that had been presented by early studies on the bacteriophages. Shortly after their discovery it was noted that some bacteriophages appeared to be "carried" by particular strains of bacteria. By carried, I mean that small amounts of the bacteriophages in question could always be found in cultures of the bacteria, while the bacteria themselves appeared to be resistant to the bacteriophages they carried. The bacteriophages involved were not defective in any way, since they were perfectly capable of infecting and killing other types of bacteria. Conversely, bacteria that carried one type of bacteriophage were sensitive to other bacteriophages. Strains of bacteria that carried particular bacteriophages came to be known as "lysogenic", since their associated bacteriophages can infect and cause lysis of other strains of bacterial cells. Until Lwoff began his experiments this phenomenon of "lysogeny" had usually been explained away as due to strains of bacteria that were contaminated with bacteriophages that they happened to be resistant to. The results of Lwoff's work demolished that notion and propelled lysogeny to the forefront of research in virology and molecular biology in general.[1]

Lwoff began his studies with a lysogenic strain of bacterium known as *Baccillus megaterium*. He chose this bacterium because it was relatively large, and he had decided to approach the problem by looking at individual bacterial cells under the microscope. This decision to examine individual cells turned out to be very fortunate, but by Lwoff's own testimony it was taken for the most unscientific of reasons: "I dislike mathematics", he wrote, "for which I am not gifted, and I wanted to avoid formulae, statistical analysis and, more generally, calculations, as far as possible."[1] The success that flowed from this initially negative attitude encourages biologists terrified of mathematics everywhere!

He added a single cell of his lysogenic bacterium to a drop of culture fluid and waited and watched. Each time the cell divided he removed one of the daughter cells, added it to a similar culture medium, and allowed it to multiply into a large bacterial colony. The remaining daughter cell was left under the microscope until it divided once more (see figure 5.1). Each of the colonies grown up from the removed cells exhibited the usual characteristics of lysogenic bacteria; in particular they always contained a few free bacteriophages which could infect and lyse other strains of bacteria. *But infectious bacteriophages could not be obtained from the culture fluid around the cell under the microscope.*

These intriguing results suggested that the ability to produce bacteriophages was being passed on to the daughter cells at each cell division. But since bacteriophages only appeared in the culture fluid of large colonies of bacteria, Lwoff felt that perhaps only a small percentage of a population of lysogenic bacterial cells might produce bacteriophages at any one time. This idea was then supported when he found that the single cell under the microscope would occasionally burst open spontaneously, after which the drop of culture fluid around it *did* contain infectious bacteriophages.

An appealing explanation for Lwoff's results was that lysogenic bacteria contain some sort of bacteriophage *precursor* which, although not itself infectious, can occasionally be activated to start

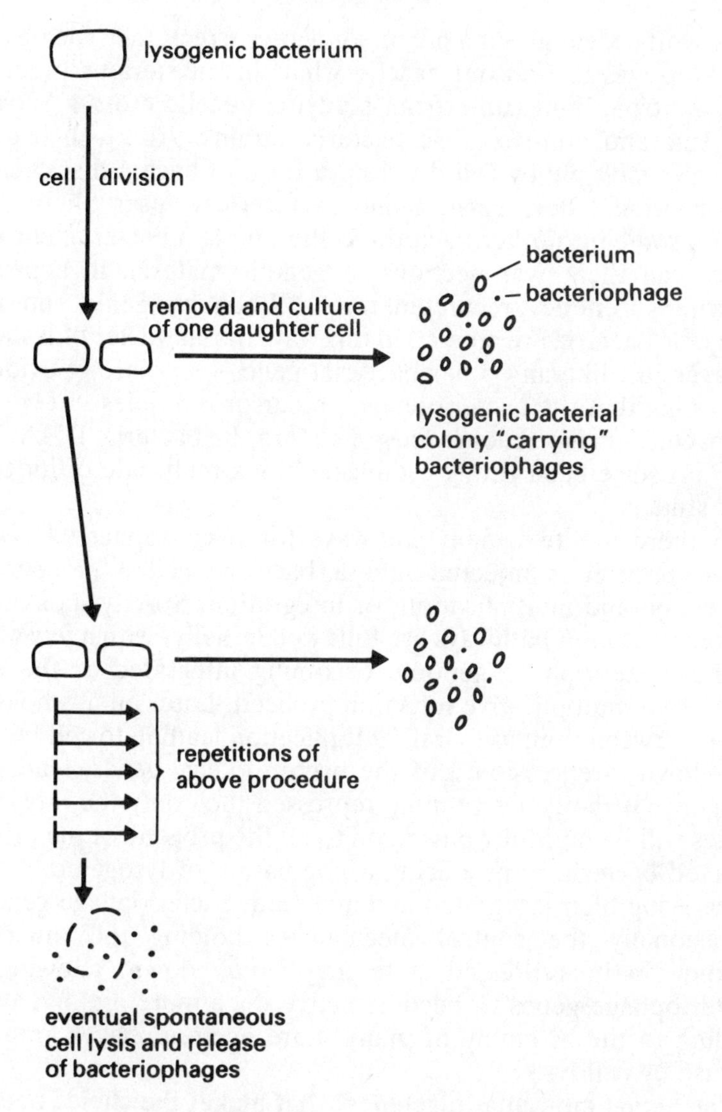

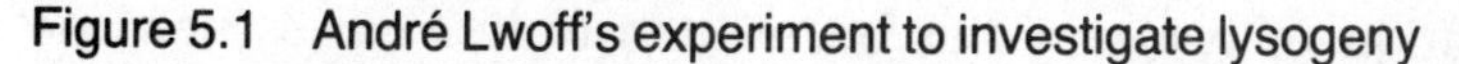
Figure 5.1 André Lwoff's experiment to investigate lysogeny

producing bacteriophages. Activation of the precursor would thus eventually result in cell lysis just like a normal multipicative infection. Having found that lysogenic bacteria would occasionally lyse spontaneously, Lwoff went on to show that lysis could also be artificially induced, for example by exposing lysogenic bacteria to ultra-violet light.

If Lwoff's ideas about a precursor were correct, then the obvious next step was to find out exactly what this mysterious precursor was. A strong hint came from studying genetic crosses between lysogenic and non-lysogenic bacterial strains. Although bacteria generally multiply by cell division, a form of bacterial sex (called "conjugation") does exist. Some bacteria can insert their single chromosome into other bacterial cells. The two bacterial chromosomes can then swap sections of genetic material in a process known as genetic "recombination". When lysogenic and non-lysogenic bacteria are crossed in this way, the character of lysogeny behaves just like any other bacterial gene – very strong evidence suggesting that the bacteriophage precursor is a quiescent copy of the bacteriophage genome *integrated* into the bacterial DNA. This idea has since been firmly established by a multitude of independent studies.

So there are two main pathways for bacteriophage DNA to follow once it is injected into a bacterial cell – independent expression and multiplication, or integration. Strictly speaking, it is likely that *both* pathways are followed initially, with a few copies of the bacteriophage genome becoming integrated as the early stages of a multiplicative infection proceed. Later on a "choice" is made between rampant viral multiplication leading to cell lysis; or shut-down ("repression") of the bacteriophage genes, leading to lysogeny. If the genes become repressed then only the integrated copies will be faithfully passed on to all the progeny of the original infected bacterium, producing a population of lysogenic bacteria all carrying their integrated and quiescent bacteriophage genome. Occasionally the control mechanisms holding the integrated genome in its repressed state may break down, allowing the bacteriophage genes to become active once more and eventually leading to the assembly of many more bacteriophages and their release by cell lysis.

The actual molecular machinery that makes the choice between multiplication or lysogeny has been intensively studied, particularly in a bacteriophage called "lambda" that infects the common gut bacterium *Escherichia coli*. It would be inappropriate to go into all the details here, but in summary the bacteriophage genome can produce the proteins required to commit the bacteriophage to *either* pathway. The eventual choice of pathway seems to result from a competition for dominance between these two sets of mutually antagonistic proteins.[2] Obviously there is great interest in what actually allows one of these two sets of proteins to become dominant. Some clues exist. For example, if bacterial cells are starved of nutrients, or if large numbers of bacteriophages are present, then the balance is tipped in favour of lysogeny. This can be rationalised as a "sensible" reaction of the bacteriophage to conditions in which the raw materials for multiplication are likely to be in short supply. The chemistry that allows bacteriophages to reach such a "sensible" decision is currently the subject of much study.

Another problem that has taxed virologists for many years is *how* does the bacteriophage genome actually manage to become integrated into the bacterial DNA? In the case of bacteriophage lambda, many details of the integration process are known, with plausible ideas available to explain the aspects that are less certain. You will find a summary of lambda's integration strategy in box 5A, and even if the details are a bit much for you it is worthwhile looking at figure 5.2 to get an overall idea of how integration proceeds.

In the meantime, having considered the integration of bacteriophage genomes, it is time to return to the viruses that infect the more complex cells of humans and other animals. There is probably no phenomenon amongst animal virus infections that directly parallels lysogeny, but animal virus genomes certainly can become integrated into cellular DNA.

Animal virus integration[4]

Some of the earliest evidence suggesting that the option of integration might also be available to animal virus genomes came from studying the "papovaviruses" – simple, non-enveloped viruses that use double-stranded DNA as their genetic material. These viruses can give us warts and possibly also cancer (abilities which are intimately linked to their ability to achieve integration). A papovavirus infection normally follows the typical multiplicative path, eventually leading to the death of the infected cells. But if mouse cells, for example, are infected with a papovavirus normally found in monkeys (called "SV40"), then the infection can follow an unusual alternative course. Instead of multiplying and killing the infected cells the virus can "transform" them into rapidly dividing tumour-like cells. These transformed cells do not produce any new virus particles, although they always contain some specific

Box 5A – THE INTEGRATION OF BACTERIOPHAGE LAMBDA DNA

The genetic material of bacteriophage lambda is injected into an *Escherichia coli* cell as a linear stretch of double-stranded DNA, containing around 40 genes. Prior to integration the DNA becomes circular due to base-pairing between short complementary single-stranded regions found at either end of the linear molecules (see figure 5.2a). The sugar-phosphate backbone of the circular DNA is sealed at the newly formed junctions by an enzyme.

Integration into the double-stranded DNA of the bacterial chromosome takes place at specific "attachment" sites shared by both bacteriophage and bacterial DNAs. Detailed analysis of these attachment sites has revealed that they each contain a central region with identical nucleotide sequences (see figure 5.2b). It is at this region of common nucleotide sequence that the integration actually takes place.[3]

The precise integration mechanism remains unclear, but obviously the presence of the common sequence means that either strand of the bacteriophage DNA in this region could base-pair with the complementary strand in the bacterial DNA. One simple mechanism by which this sort of strand exchange could lead to integration is shown in figure 5.2. If staggered breaks were introduced into both DNAs at identical sites (see figure 5.2b), then the cut ends of the bacteriophage DNA could become base-paired to the complementary cut ends of the bacterial DNA (see figure 5.2c). Enzymes could then reseal the sugar-phosphate backbones of the DNA molecules, leaving the bacteriophage DNA integrated into the bacterial chromosome.

The evidence in favour of this general integration mechanism is very strong. Both bacteriophage and bacterial enzymes are required to catalyse integration. The exact role of these enzymes is not known, but they can bind to the DNA at the attachment sites and the bacteriophage enzyme can break and reseal the sugar-phosphate backbone of one strand of a DNA double-helix. These activities make good sense in the light of the overall mechanism considered above.

Excision of the integrated bacteriophage genome can also occur, for example, when the repression that maintains lysogeny breaks down. The freed genome then initiates bacteriophage multiplication.

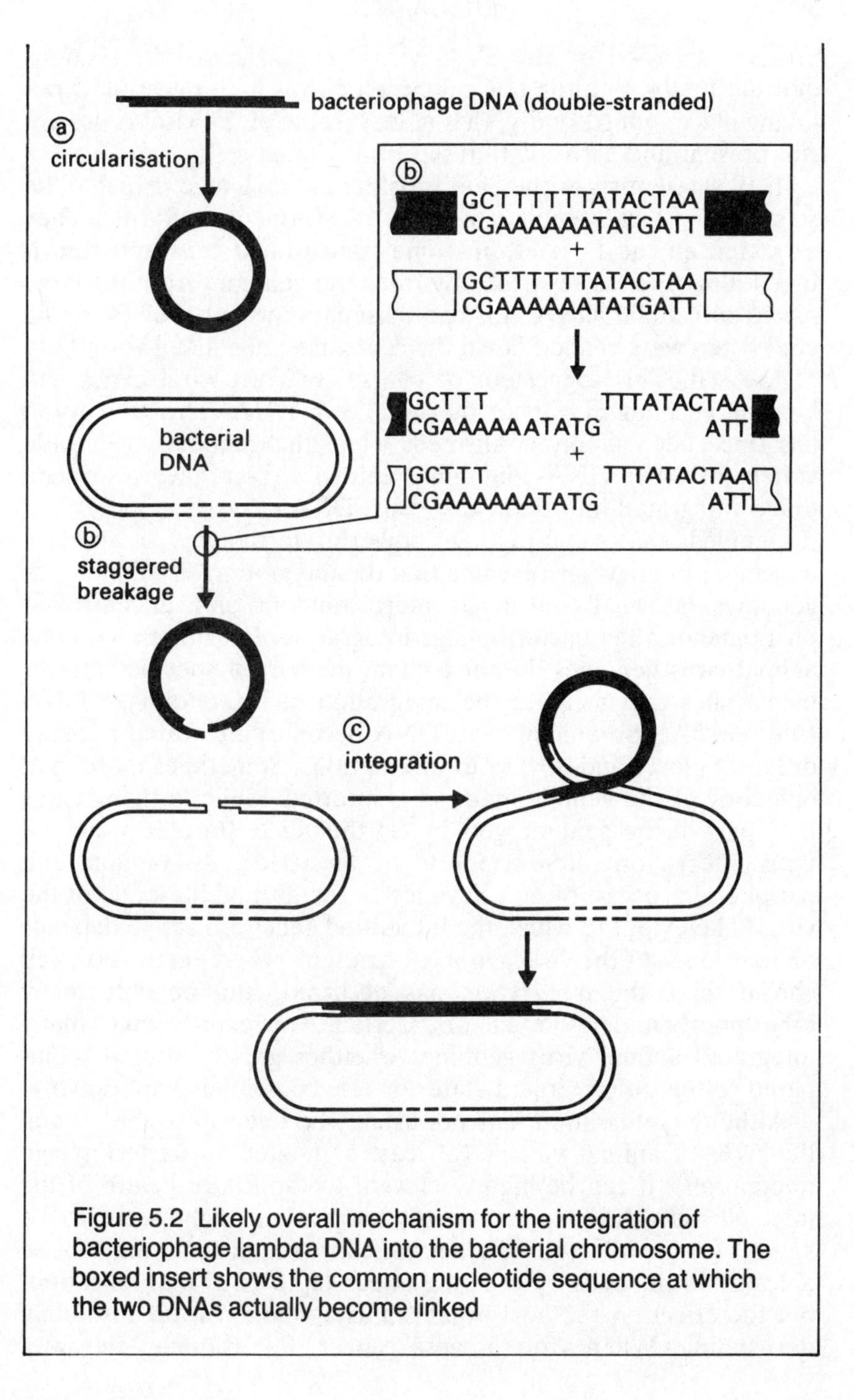

Figure 5.2 Likely overall mechanism for the integration of bacteriophage lambda DNA into the bacterial chromosome. The boxed insert shows the common nucleotide sequence at which the two DNAs actually become linked

proteins encoded by the SV40 virus genes. These observations indicate firstly, that the normal cycle of virus multiplication is not taking place, but secondly, that at least some of the viral genes are still present and active within the transformed cells.

In 1968 scientists at the Salk Institute in San Diego revealed the state of the viral genome in cells transformed by SV40.[5] They extracted all the DNA from some transformed cells and centrifuged it in order to separate any free viral genomes from the large pieces of cellular DNA. Instead of separating out, however, the viral genes were carried down the centrifuge tube along with those of the cell. This suggested, of course, that the viral genes had become an integral part of the cell's own DNA. This conclusion has since been amply confirmed, while the genomes of a wide variety of other DNA-containing animal viruses have also been found integrated into their host cells' DNA.

Detailed analysis using the powerful techniques of modern molecular biology has revealed that the integration of animal virus genomes is usually a much more random and uncontrolled phenomenon than bacteriophage integration. For one thing, most animal virus genomes do not contain the sort of specific "attachment" sites that mediate the integration of bacteriophage DNA (see box 5A). So animal virus DNA becomes integrated at many different places and in various orientations. Sometimes more than one copy of the genetic material is inserted, while in many cases only part of the genome gets in. So the whole process of animal virus integration often seems to be an accidental, random and peripheral process, of no relevance to the normal life-cycle of the virus. The extent to which the integrated genes are active depends on how much of the viral genome is present, where on the host cell chromosome the integration has occurred, and possibly many other poorly understood factors. Certainly the expression of many integrated animal virus genomes is either greatly altered (compared to the un-integrated state) or else completely shut down.

Although integration may not usually be relevant to the normal life-cycle of animal viruses (at least compared to bacteriophage integration), it can be highly relevant to the future health of the infected individual. In one sense integration might seem to be "good" for the infected cells, since the integrated genes may not be able to kill the cells by bringing about rapid virus multiplication, but the effect on the host organism as a whole can be absolutely devastating. When viruses cause cancer, for example, the integration of the viral genome seems to be a crucial step towards the onset of the disease (see chapter 10). This fits with the earlier observation that integration of the SV40 virus genome can transform mouse cells into rapidly dividing "tumour" cells. Cancer is probably not the only problem precipitated by viral integration, for some extremely troublesome "persistent" virus diseases that can linger on for years may be maintained by the continual presence of integrated viral DNA (see chapter 9).

We shall obviously return to the subject of animal virus integration when we come to consider the illnesses that viruses can cause, but the major message in the meantime is that it is typically a haphazard and poorly understood phenomenon that has little to do with viral multiplication. For one very special and fascinating type of animal virus, however, precisely the reverse is true. These are the "retroviruses", for whom the process of integration is a central aspect of the normal cycle of multiplicative infection. Paradoxically, retrovirus genes enter the cell in the form of single-stranded RNA, despite the fact that they must obviously be copied into double-stranded DNA before being integrated into the double-stranded DNA of the cell's own genome. Things are rarely straightforward with the viruses!

Retroviruses[4,6,7]

The most important illness caused by the retroviruses is cancer, and their ability to cause cancer is a direct result of their intimate relationship with the genetic material of the infected cell (see chapter 10). When a retrovirus enters a cell an enzyme carried by the virus soon converts its single-stranded RNA genome into a double-stranded DNA copy (see figure 5.3). This vital retrovirus enzyme is known as "reverse transcriptase", since it reverses the normal process of transcription in which double-stranded DNA is used to produce mRNA. Once copied into double-stranded DNA, the retrovirus genes integrate into the host cell DNA, and it is in this integrated form that they actually become active in viral multiplication. So both the viral mRNA used to produce proteins, and the RNA copies of the genetic material needed for incorporation into new viruses, are copied from an integrated DNA template.

In contrast to the integration of most animal virus genomes, retrovirus integration seems to be a highly efficient and carefully

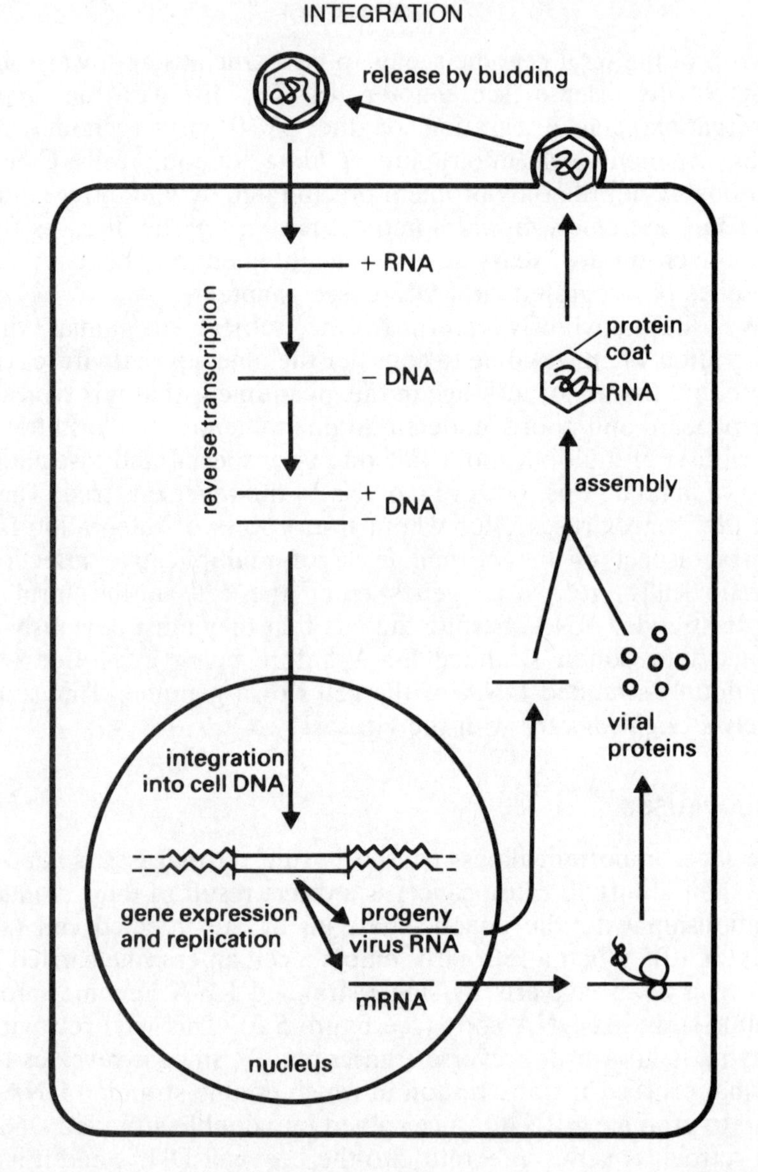

Figure 5.3 Life-cycle of the retroviruses

controlled event. Specific regions in the viral genome always form the boundaries between viral and cellular DNA, although the integration can occur at many different sites on the host cell chromosome. The question of whether or not integration is restricted to specific sites (albeit a large number of them) is still open; and the precise mechanism of retroviral integration remains unclear.

The eventual release of newly made retroviruses is achieved by budding from the host cell membrane – a non-destructive exit path that often allows retroviral infections to be tolerated for long periods of time. In fact, despite their famed ability to cause cancer, many retroviral infections are apparently harmless.

Since they routinely link up with host cell chromosomes, it is not surprising that the integrated genomes of retroviruses can often behave much like the normal genes of the cell.[8] Of particular interest is their ability to be passed down through succeeding generations of animals just like any other inheritable trait. This inheritance of retrovirus genomes presumably stems from the occasional infection of an animal's sperm or egg-producing cells (the "germ" cells). Every one of us might well contain such inherited retroviral genes joined up to our normal "human" DNA.[9] The presence of such "viruses within" probably should not worry us too much, for in most cases studied so far they seem to do their hosts no harm. Sometimes they may remain completely inactive, presumably held in check by the rigid control systems governing the activity of all cellular genes. In other cases they are only partially active (producing only one viral protein, for example), while occasionally fully infectious viruses can be produced. Inactive inherited retrovirus genomes can often be "induced" to become active by electromagnetic radiation or various chemical carcinogens. In a few cases such activation has been linked to the later development of cancer in laboratory animals. As we shall see in chapter 10, the whole question of the links between retroviruses and cancer has recently been one of the most active and fruitful areas of enquiry in all biology.

So in summary then, the act of biological piracy that we call a viral infection does not always follow the simple multiplicative path described in chapter 4. One of the most dramatic variations involves the integration of the viral genome into the host cell's own DNA; an event that can lead to a complete or partial shut-down of viral gene activity, virus multiplication as before, or the committal of the host cell to the cycle of uncontrolled growth that leads to cancer. The two "main streams" of viral activity summarised in the last two chapters by no means exhaust all of the possibilities available to these intriguing infectious agents. We shall meet some further variations later.

But are they "alive"?

This is the most common question asked by laymen and students on their first introduction to the viruses. It is a question that featured prominently in the early days of virology, and which to some extent is still debated today. It is not uncommon to hear one biologist chastising another for referring to the viruses as "microorganisms", but the debate has largely been abandoned as an irrelevance. The most appropriate response to the question is firstly, it depends on what you mean by "alive", and secondly, does it really matter in any case?

The most valuable outcome of any discussion on this subject is that it reveals what vague terms "alive" and "life" really are. Initially, many people feel that there must surely be some clear-cut and obvious difference between the live and the inanimate. But such confidence stems from examining nature at its *extremes*. Few would contend that rocks and stones are alive; while all can agree that they themselves, their pet rabbit and even the smallest of insects most certainly are alive. The problem comes when you descend the ladder of complexity, past the plants and fungi, protozoa, bacteria, and down to and beyond the viruses themselves. Where do you draw the line?

Some textbooks will provide you with a list of "vital characteristics" of living things, such as locomotion, nutrition, growth, respiration, excretion, sensitivity and reproduction. If a "common-sense" interpretation of such a list is to be the criterion then the viruses are not alive. But the problem with such lists is that the terms used are themselves open to different interpretations. To what extent can plants be said to display locomotion, for example; to what extent are bacteria "sensitive", when compared to the exquisite sensitivity of the human nervous system? Can the growth of a crystal from molecules in solution be considered to involve growth, nutrition, reproduction and locomotion? Crystal formation is certainly "sensitive" to changes in the surrounding chemical environment. In any case, what justification can we give for deciding on such a list, other than the attempt to standardise our instinctive "feel" for what is dead and what is alive?

More modern textbooks often declare that in order to be classified as "alive", something must simply be able to *evolve* by natural selection. In other words it must be able to make copies of itself and also to gradually change. Changes that allow it to make even more copies of itself will then be retained, simply by bestowing a reproductive advantage on the altered form. The justification for making the ability to evolve the principal criterion for life, is that it is this ability that has allowed parts of the inanimate primeval world to develop into the clearly accepted "living" creatures of today. So the capacity for evolution must be the first step on the ladder of life, and by this more modern criterion the viruses most certainly are alive.

As a third alternative, some textbooks might tell you that the fundamental characteristic of living things is their apparent ability to defy the second law of thermodynamics by assembling complex organised biological structures from the disordered non-living world. Of course thermodynamic laws remain inviolate overall, with the driving force for life's relatively feeble movement in reverse being supplied by the massive flow of energy from the sun. Thermodynamic criteria certainly grant the status of life to the viruses and other entities that seem mere chemicals according to more "common-sense" definitions. A single strand of nucleic acid in the primordial soup, for example, by acting as a template to make complementary strands of itself, would qualify to be called "alive". But then what about a fridge that cools its interior in apparent violation of the second law (but simply because it is supplied with electrical energy through the flex to power its "defiance"). Is a fridge alive just as we are, with the energy of the sun equivalent to the electrical energy brought through the flex?

To focus more closely on the viruses, some people deny them the status of life-forms since they are clearly not independent "free-living" creatures. But then neither are we. Just as the viruses are dependent on the metabolism of their host cells, so we ourselves are dependent on the complex chemistry of plants to trap solar energy and provide food for ourselves or our livestock. Humans are certainly less obviously dependent on other forms of life than the viruses are, but the dependency is there nevertheless.

I am not trying to supply any rigorous analysis of the alternative definitions of life here; simply giving you a sample of the various points that combatants in the debate fire off at one another. What clearly emerges is that life is certainly not an easy term to define. The concept, not of individual life-forms, but of "life" as a massive integrated biological system spread across the globe, is in many ways much more satisfactory. Within such a system the viruses are certainly *a part of life*, just as we ourselves are a part of life. We are certainly a more independent part than the viruses, but still not completely independent from the other parts of the biosphere. But there still remain problems in deciding where the boundaries of this united biosphere should be set.

So faced with all the difficulties in deciding what we really mean by "life", the response of "does it really matter?" becomes increasingly attractive. Perhaps it is best simply to recognise the inadequacy of the term "life" and concentrate our thoughts on *what happens* in the world we find around us, rather than fretting over classifications and definitions. The stars, the rocks, the viruses and "higher" organisms such as ourselves all exist and interact. It is interactions and changes that really matter, not definitions.

For the viruses to interact with the cells of our body they must first find a route from the environment outside into the vulnerable cells within. We must now turn our attention to the refreshingly practical question of how such viral invasion can come about.

CHAPTER 6

Invasion – strategies for a gatecrasher

The horrific photograph shown in plate 6.1 justifies the anxiety we all feel when trapped beside a cold-ridden "sneezer" on the bus, tube or train. It is a perfect evocation of those aspects of mankind's interaction with the viruses that cause us most concern. We are interested in the viruses mainly because they make us ill. If they did not make us ill then you probably would not want to read an entire book about them and it would probably have taken much longer to discover them in the first place. But before they can cause any illnesses the viruses have a formidable array of obstacles to surmount. First of all they must get *onto* us, then they need to get *into* us, and finally they need to meet up with a type of cell that they can infect and successfully multiply within. The whole process often begins with the person next to us behaving like the gentleman in plate 6.1, and then ends with our own version of the same miserable performance.

Any assistance given to the viruses by an unmuffled sneeze is certainly of benefit to them. After all, the inactivity of viruses outside of cells means that to spread from host to host they are dependent on being carried along by secondary agencies such as wind and water. Alternatively, of course, they might remain where they come to rest after release from an infected individual, dormant until someone else arrives to pick them up. They certainly cannot fly, run, jump or swim like many more complex parasites. So the subject for this chapter is the way in which "invading" viruses can overcome these disadvantages to reach the vulnerable tissues and cells of their hosts.[1,2]

You will find a summary of the major sites at which viruses can

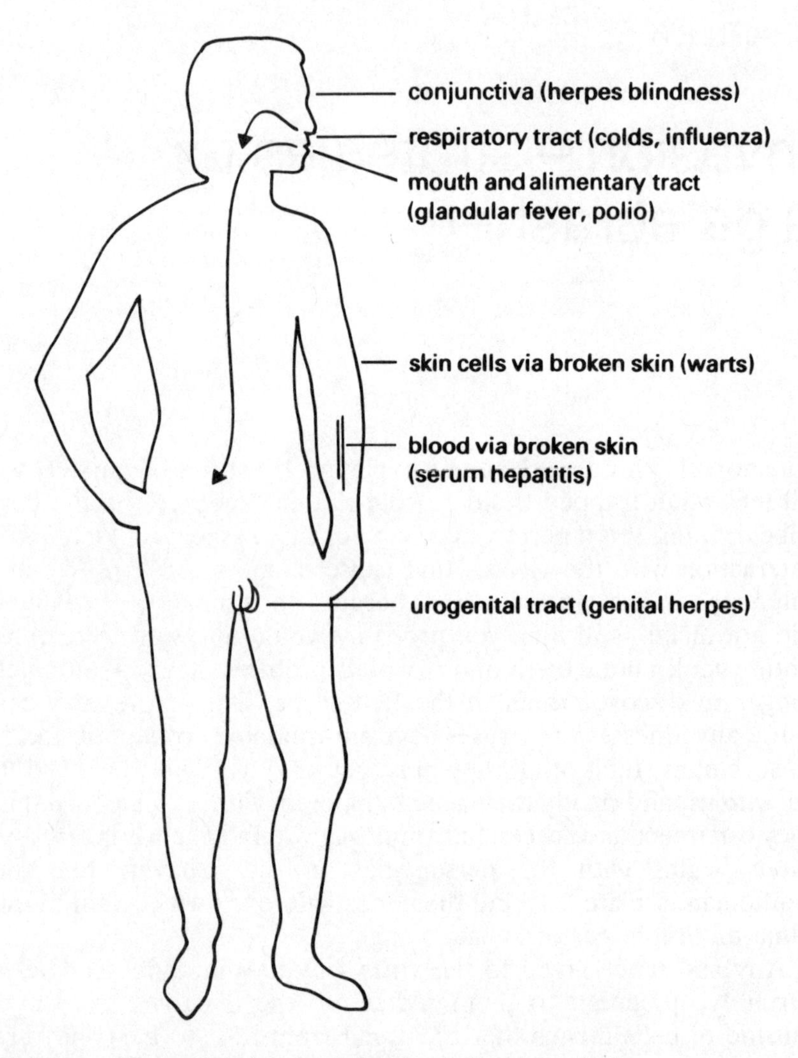

Figure 6.1 Entry sites for viral infection (with examples)

initially invade the human body in figure 6.1, while figure 6.2 portrays the main ways in which viruses can travel between these sites. By far the commonest mode of viral transmission is also the one with which people are most familiar – the link between our respiratory systems mediated by virus-laden aerosols. Such aerosols are produced by coughs and sneezes, but also arise in the course of normal breathing. They provide transport not only for the familiar respiratory infections such as the cold and influenza,

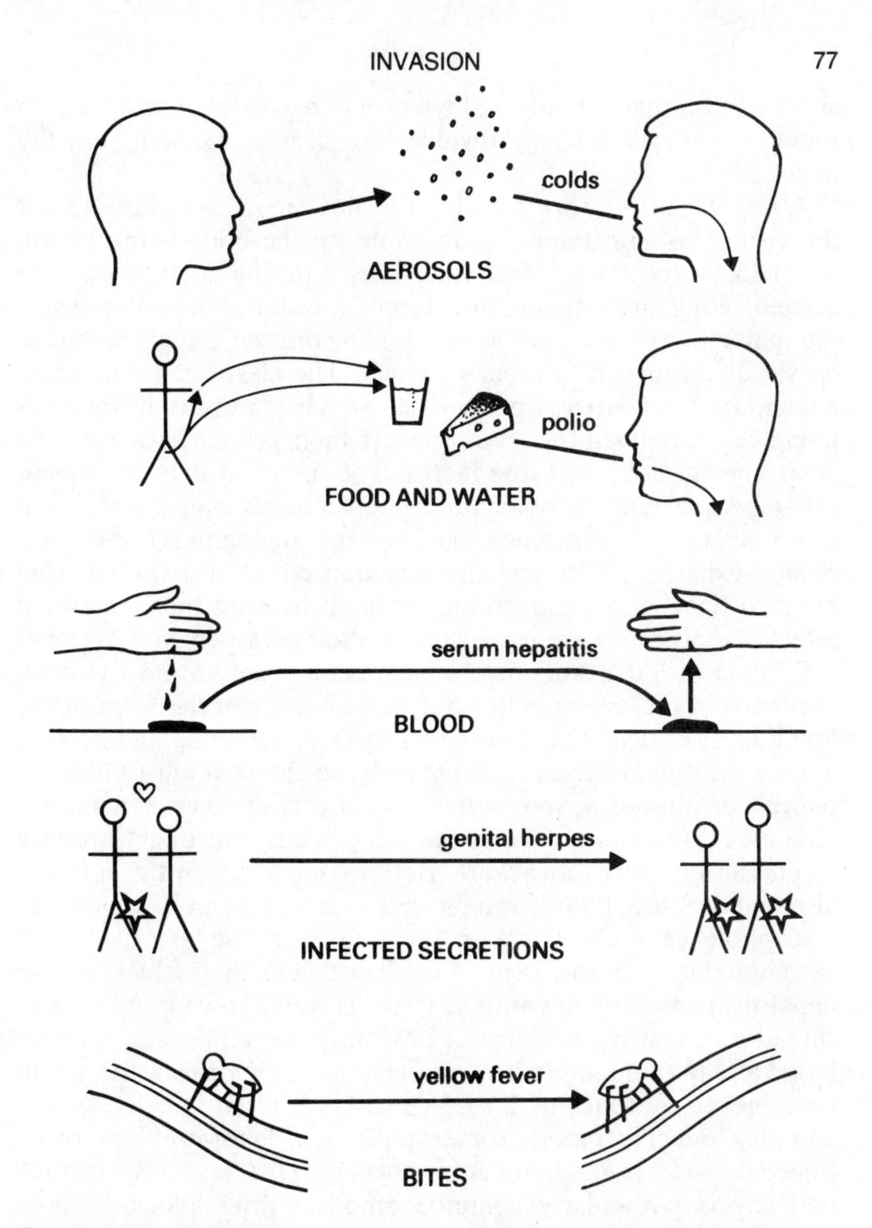

Figure 6.2 Common means of virus transmission (with examples)

but also for many other diseases including chickenpox, mumps and measles. When the viruses responsible for these latter diseases reach the respiratory system they use the cells they find there as merely a first staging post on the way to other parts of the body. The infected droplets carrying all these viruses can remain airborne for many hours, allowing a virus to be passed on to someone who has had no obvious contact with the source of the infection.

After the respiratory system, the next most popular site for the viruses to gain their first foothold on the body is the mouth and alimentary tract. Most infections entering in this way are carried along in contaminated food or water. Glandular fever, though, with its reputation as the "kissing disease", might well pass on via direct mouth to mouth contact. The classic example of an alimentary tract virus is polio virus, which multiplies in the cells lining the throat and intestine and is then passed back into the environment along with the faeces. The infection then spreads to other people who drink appropriately contaminated water – a common enough occurrence when sewage treatment is either poor or non-existent. Of course the infection could also spread from hand to hand and then mouth, or hand to food then mouth, if infected individuals are careless with their own personal hygiene.

Contact with the skin must be a common event for most viruses, but the skin presents a major defensive barrier in the form of the tough layer of dead skin cells on its surface. To set up an infection a virus must gain access to *living* cells, so the skin must either be pierced or injured in some other way if it is to serve as a site for viral entry. Infection of living skin cells themselves might produce a generalised "sore" or a wart. But of course cuts in the skin can also expose the blood, which can either transport viruses to susceptible cells elsewhere in the body, or else provide direct accommodation in the form of the blood cells themselves. Serum hepatitis, caused by hepatitis B virus, is well known to be able to spread in this way. A cut must obviously come into contact with blood already contaminated with the hepatitis B virus, which might be achieved by touching a surface carrying dried infected blood, handling infected blood (for example in a laboratory) or being injected with a contaminated needle. This last transmission pathway is particularly common amongst drug addicts, whose standards of hygiene are generally much poorer than those found in hospitals (where sterilised needles are usually used once and then discarded). Donated blood is routinely screened to avoid transmitting serum hepatitis along with a life-saving transfusion.

Viruses can also penetrate through the skin via the bite of an infected animal. The dreaded rabies virus, carried along with the saliva of dogs, is one example; while yellow fever virus, which kills one in ten of its victims, is passed into a human by the bite of an infected mosquito.

The urogenital system is actually a relatively uncommon site for viral infections, despite being the most talked about in recent years thanks to the media attention given to genital herpes. Referred to by most people simply as "herpes", this disease is caused by specific members of the herpesvirus group known as the herpes simplex viruses. (Other herpesviruses produce such varied illnesses as chickenpox, shingles, glandular fever and encephalitis.) The herpes simplex viruses can multiply in the tissues of male and female genitalia, being transmitted by direct contact with infected mucous secretions during intercourse.

The final site of viral entry listed in figure 6.1 is the conjunctiva – the layer of living cells that forms the outer covering of the eye. This is a relatively unimportant site of infection compared to the others considered above, but the consequences of infection can be extremely important to the individuals concerned. One of the commonest infectious causes of blindness, for example, is conjunctival infection with herpesviruses.

This brief look at the sites at which viruses can gain entry to the body presents an obvious question: once inside the body, why do most viruses infect only specific types of tissue while they usually come into contact with a variety of tissues? Rhinoviruses, for example, infect the cells lining the nose and throat to cause colds, but they must also get washed down into the stomach and the intestine. Why do the rhinoviruses not infect the cells lining the alimentary tract while other viruses, such as poliovirus, certainly do? A complete answer to fit every case is certainly not available, but two crucial factors seem to be involved: firstly, the presence of suitable cell-surface receptors for the viruses to bind to, and secondly, the existence of various "barrier" systems that can obstruct viral transport into and throughout the body. A discussion of receptors and barriers will not only throw more light on the initial sites of viral infections, but will also lead us into the crucial subject of how an initial infection can later *spread* to many other parts of the body.

Receptors and barriers

I have already told you that viruses must bind to specific receptor molecules on the cell surface before they can enter a cell, so the importance of these receptors in determining which viruses enter which types of cell should be obvious. Clearly, if a cell lacks a suitable receptor for any particular virus then that virus is not going to be able to get inside the cell. Of course in the absence of any experimental evidence it is conceivable that suitable receptors might be found on all types of cell. But experiments to resolve this issue *have* been performed, and they suggest that the cell and tissue specificity of a virus can indeed be determined by the availability of receptors. Poliovirus, for example, only infects man and the other primates, apparently because non-primate cells lack suitable receptors. If you purify the poliovirus genome and artificially insert it into a non-primate cell, then the genes soon become active and more polioviruses are produced. So there is nothing about the internal chemistry of the non-primate cells that prevents poliovirus multiplication, but their lack of receptors makes it impossible for the virus to get inside the cells in the first place.

Such experiments tell us that the range of different *species* infected by poliovirus can be determined by receptor availability, but what about the range of different *cell types* that become infected within one particular animal? Poliovirus enters through the mouth and then multiplies in the throat and intestine before being passed out of the body along with the faeces. But the infection need not be restricted to the alimentary tract. Like many other viruses, poliovirus can spread to other parts of the body after a brief period of multiplication in the cells that are first infected. In the case of poliovirus, the viruses can eventually find their way into the bloodstream, which then carries them to all parts of the body. But despite being widely circulated, the viruses only infect specific types of cell – particularly some nerve cells found in the spinal cord. It is the infection of these nerve cells that can cause the paralysis associated with "polio". Many of the other tissues that must come into contact with poliovirus have been found to lack receptors for the virus to bind to. So the tissue specificity of poliovirus is at least partially explained by receptor availability.

A second major influence on the types of cells, and therefore tissues, infected by particular viruses, is the presence of so-called "barrier systems" which can exclude viruses from particular sites. The most obvious barrier is the skin, which physically prevents a virus from reaching the vulnerable living cells beneath it. But more subtle and selective barriers are erected by the varying chemical conditions found throughout the body. The cells lining the respiratory and urogenital systems, for example, secrete a sticky fluid called "mucus" which can act as an effective barrier to many viruses. In some cases the viruses may be hindered simply by the physical presence of the mucus, but it also contains antibodies and other proteins that can bind to particular viruses and so prevent them from entering cells. Of course mucus is certainly not a perfect barrier, since many viruses *can* infect the cells protected by a mucus lining. In some cases these infections probably begin at regions of injury, where the mucus barrier is incomplete; but some viruses (such as influenza virus) carry enzymes on their surface that might actually "eat" their way through the mucus to the living cells below.

The acidic environment of the stomach is another major barrier system. It can inactivate many viruses washed down from the mouth and throat and so prevent the infections they cause from spreading to the digestive system. The vulnerability of the rhinoviruses to acid at least partially explains our earlier observation that these viruses can cause colds in the respiratory system, but do not infect the stomach or the intestine. Obviously poliovirus, on the other hand, must be able to withstand the stomach's acidity to pass on and initiate infection when it reaches the intestine.

I could continue with details of other barrier systems, such as the specialised membranes of the placenta, the cells lining the blood capillaries of the brain or even differences in temperature between different parts of the body. But we are more concerned with general principles than great detail here; and the overall general principle is that the body can be divided into many distinct compartments, each with their own characteristic chemical environment and protected by different types of membranes or secretions. Since different viruses vary in chemical composition, size, enzymic activities and so on, their ability to enter the various parts of the body also varies. The need to surmount the barriers, and then to find suitable receptors on the cells beyond, can probably explain the tissue specifity of most viruses.

Spread[3]

As we have already observed with poliovirus, an infection is certainly not restricted to its initial site of entry and multiplication.

Infection of the various entry sites illustrated in figure 6.1 is frequently followed by a more widespread distribution of the viruses throughout the body. We have looked at some of the factors determining what types of cell might be vulnerable to a spreading infection, but what routes are available by which the spread can take place?

There are actually five main routes available, summarised in figure 6.3. Perhaps the most obvious possibility is for an infection to simply spread from cell to cell. As soon as a new crop of viruses is released from any initially infected cell, then they will be able to infect any similar types of cell nearby. Alternatively, the viral genetic material might pass through the tiny channels that link up the cytoplasm of adjoining cells. The advantage of this route is that free virus particles do not have to brave the hostile world outside the cells, where they are more vulnerable to antibodies and the other agents of the body's defence systems. Measles virus solves the problem of cell-to-cell spreading in another way which also avoids the release of free viruses to the extracellular world. Like some other viruses, measles virus can cause the membranes of infected cells to fuse with those of neighbouring cells, producing giant infected "cells" containing many cell nuclei.

These various routes for an infection to spread to neighbouring cells explain how a large area of infection can develop wherever a virus first enters the body; but they do not tell us how the infection can spread throughout the whole body, or at least to several distinct organs. The most flexible system available to mediate such widespread dissemination is the circulatory system, which can transport viruses along with blood or lymph. Looking through the translucent layer of skin that covers our bodies, all we can make out of the circulatory system are a few of the larger veins. But the majority of blood vessels are tiny capillaries forming a fine three-dimensional network that takes nutrients and oxygen to every cell in the body, and then carries the waste products of cell metabolism away. The blood is brought from the heart to the capillary network at high pressure along the arteries, and having passed through the capillaries it is then collected and returned to the heart by the veins. In addition to these blood-carrying vessels, there is a separate branch of the circulatory system that transports a watery fluid called "lymph". This is the fluid that seeps out through capillary walls and into the spaces around cells. The lymph vessels drain the lymph back into the veins. So blood and lymph can carry the viruses to all parts of the body, transporting them either free in the moving fluid or enclosed within infected blood cells.

The importance of the circulatory system to the spread of viral

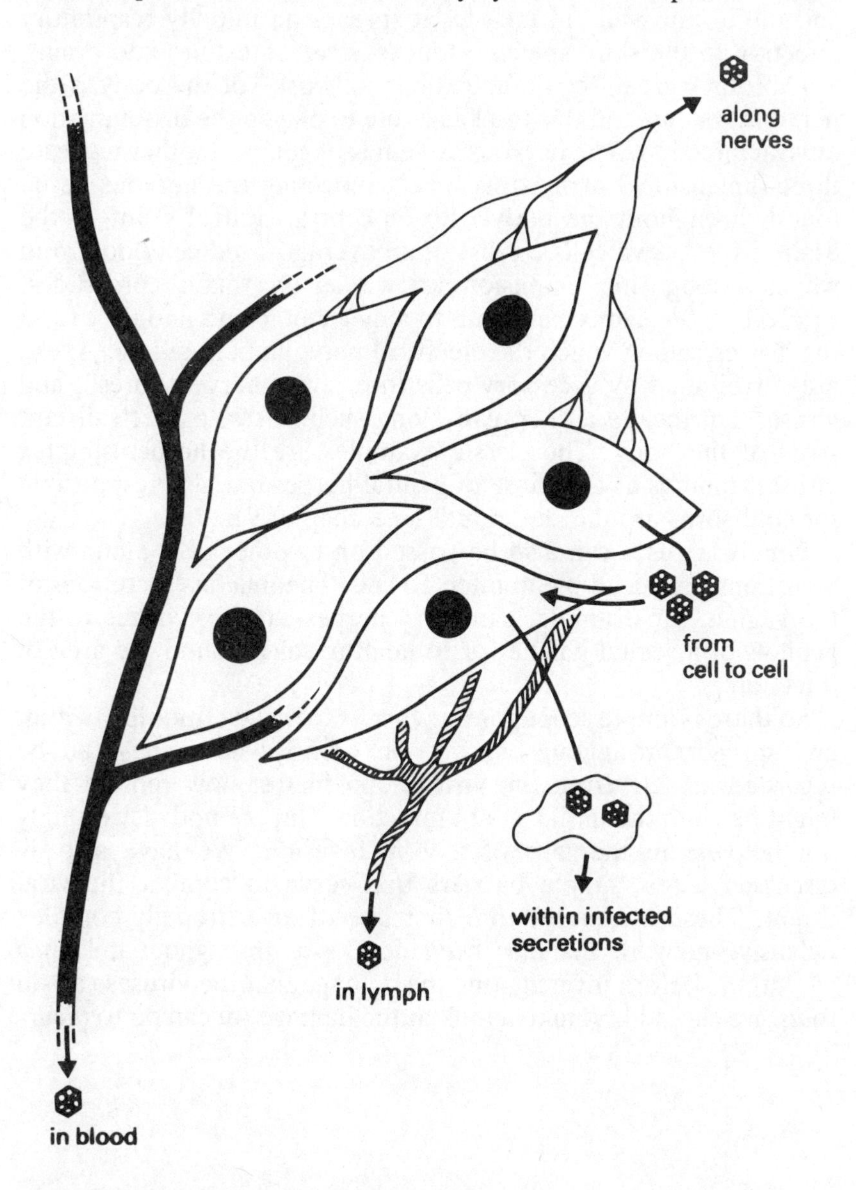

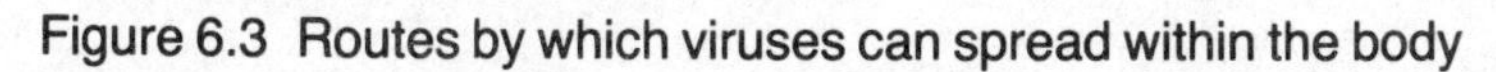
Figure 6.3 Routes by which viruses can spread within the body

infections cannot be over-emphasised. We have already seen how it can carry polioviruses to nerve cells where they can induce paralysis. In mumps, it can carry virus from the respiratory system to such organs as the salivary glands, testes or ovaries, pancreas and the brain; while in measles it spreads an initially respiratory infection to the skin, spleen, kidneys, liver, intestines and brain.

The other great "communications network" of the body is the nervous system, and this too has a role to play in the dissemination of viral infection. The nervous system is of course another intricate three-dimensional array, this time connecting the nervous tissue found throughout the body with its central control point — the brain. Most nerve cells consist of an average-sized cell body from which a long thin extension known as an "axon" protrudes. Incredibly, the axons can be up to a metre in length and they form the "wires" along which the electrical nerve impulses flow. Axons are surrounded by accessory cells to produce nerve "fibres", and viruses can invade and spread along such fibres to reach distant parts of the body. The classic examples are the herpes simplex viruses, famous as the cause of genital herpes but also responsible for cold sores around the mouth (see chapter 9).

Finally, viruses can also be passed on to other cells along with secretions produced by an infected cell. The mucous secretions of the vagina, for example, can carry herpes simplex viruses to the genitals of a sexual partner or to healthy cells around the area of infection.

So there is ample scope for the viruses to invade, multiply within and spread throughout our bodies. Clearly, no tissues can be regarded as safe from the viruses, no matter how remote they might be from the initial site of infection. But the body is certainly not helpless in the face of a viral invasion. We have already discussed a few simple barriers that serve to combat the viral threat. These are merely the first line of an extremely complex defensive network that has been developed throughout millennia evolution. Before investigating the damage that the viruses can do to *us*, we should first take a look at the damage we can do to *them*.

CHAPTER 7

Defence – the body fights back

Our relationship with the viruses and other microbes is clearly an ancient one, stretching back far beyond the point at which mankind emerged from its non-human origins. One of the legacies of this age-old struggle is the marvellously intricate and effective defensive network known as the "immune system".

Most people, when asked about immunity, would relate a tale of "antibodies" and our ability to quickly fight off infections to which we have previously been exposed. Strictly speaking, however, immunity is a universal term covering all of the body's defences. It includes both "non-specific" defences (i.e. those active against a broad range of different viruses) such as the barrier systems already examined; and also the "specific" defences which are activated against *specific* viruses, to which we may then become "immune". The story of these specific defences involves many more characters than the celebrated antibodies, which actually play a secondary (albeit important) role in the defence against viral infection.

Apart from the simple barriers, the main components of our immune defences are to be found in the circulatory system – a logical site since it reaches all parts of the body. Blood is actually one of our largest organs, accounting for about 8 per cent of body weight and made up of a baffling assortment of different cells. Some degree of simplicity is introduced by the fact that all blood cells are derived from common precursor or "stem" cells found in the bone marrow. These stem cells develop into diverse types of "white" blood cells, in addition to the red blood cells that transport oxygen, and the tiny cell fragments known as "platelets". It is the white blood cells that are responsible for immunity.

Phagocytosis[1]

Assuming that an invading virus has not infected the individual concerned before,the first defences it will confront are likely to be non-specific – directed at all viruses and not just the one causing the current bout of infection. After the barrier systems, non-specific immunity is next mediated by various types of white blood cell that can engulf and digest invading viruses – a process known as "phagocytosis". The nomenclature can be kept simple by referring to all these white blood cells as "phagocytes". They can be regarded as simply scavenger cells, ready to eat up any foreign material they encounter. Many of them, however, perform various tasks other than phagocytosis.

Like other immune cells, the phagocytes are by no means restricted to the blood. They can pass out from blood vessels into the surrounding tissue, and they also find their way in large numbers into the lymphatic system and then back to the blood. So wherever a virus invades there should be phagocytes waiting to provide an early and very effective challenge to the infection. In the early hours after infection the role of the phagocytes is paramount since, unlike the cells that mediate specific immunity, they can act without delay. In evolutionary terms phagocytosis is one of the most primitive means of defence available to multi-cellular organisms, but it is also one of the most effective. It seems likely that many viral invasions might never get past the phagocytes to trouble our more complex secondary defences.

Natural killer cells

One of the problems in discussing the various medical aspects of virology is that in many cases our knowledge is far from complete. There are still plenty of things that are not known about how viruses interact with individual cells, but even more mysteries arise when we turn to their relationship with entire organisms. One particular current puzzle concerns the role of an intriguing class of white blood cells discovered in the 1970s and known as "natural killer cells".[2] As their name suggests, these cells can apparently kill cells that have become infected with viruses, and the label "natural" indicates that they are effective immediately after infection against a wide range of different infections. This distinguishes them from the cells of the specific immune response, which as we shall soon see must be *activated* by the binding of viruses to specific receptors on the cell surface. Natural killer cells are currently the subject of much research, not only because of their possible role in fighting viral disease, but also because they may be able to kill cancer cells as soon as they arise within a healthy person. Whether or not they play a crucial role in fighting off viruses remains to be seen, but they are certainly worth a mention since they may eventually turn out to be a major part of our anti-viral defences. And remember, they work by killing *host cells* that have become infected with viruses, not by attacking the free viruses directly. So they work in a completely different way from the phagocytes.

Interferon

Moving quickly on from the still arcane world of natural killer cells, we come to an anti-viral protein that everyone must have heard about – "interferon". Although first discovered in the 1950s, interferon has probably been the most celebrated chemical of the early 1980s. It was discovered as a result of its anti-viral activities, leading to great interest in its potential as an anti-viral drug, but until recently it had proved impossible to obtain the quantities of pure interferon needed for rigorous clinical trials. That problem has been dramatically eliminated by the genetic engineering revolution, which has allowed the interferon gene to be inserted into yeasts or bacteria, which then multiply and produce plentiful amounts of the precious protein. Trials of interferon in the treatment of both viral infection and cancer have now become commonplace (see chapter 12), but it is still too early to predict how useful it will eventually turn out to be. But how does it work?

The first thing to say about interferon is that the term really refers not to one specific protein, but to a large number of closely related proteins. These "interferons" are produced by many different types of cell (perhaps all types) when they become infected with viruses. The interferon molecules are then secreted from the infected cells, allowing them to bind to the surface of neighbouring healthy cells. This binding of interferon to a cell converts it into an "anti-viral state" in which the onset of viral infection is strongly inhibited (see figure 7.1). So the production and release of interferon allows the infection of one cell to bring about the protection of many neighbouring cells against infection.

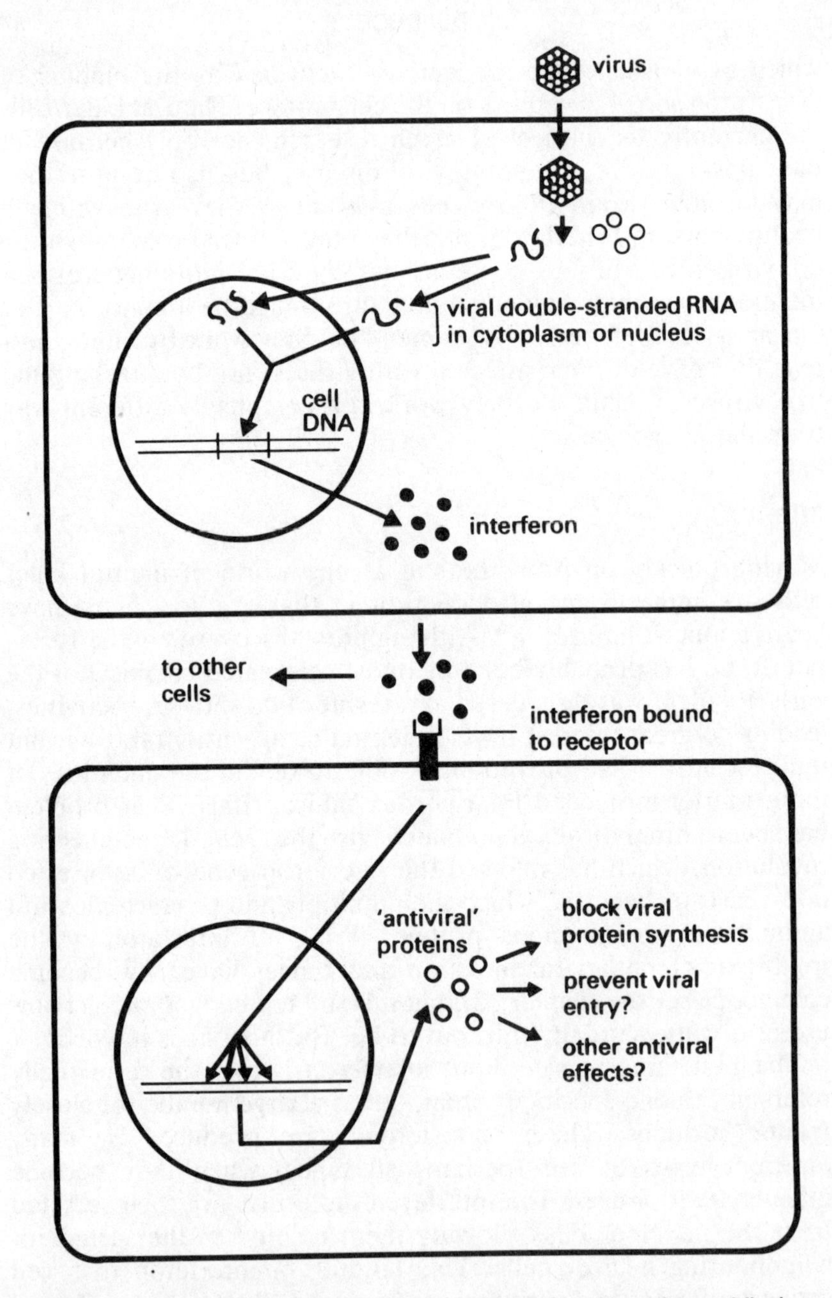

Figure 7.1 Known aspects of the anti-viral activity of interferon. Viral double-stranded RNA induces interferon production in infected cells (top). The interferon then binds to neighbouring cells, stimulating the manufacture of the proteins responsible for the "anti-viral state" (bottom)

The main details worked out so far about interferon's action are summarised in figure 7.1. One major signal that stimulates an infected cell to make interferon appears to be viral double-stranded RNA. Double-stranded RNA is formed during the multiplicative cycle of many (perhaps all) viruses, probably including at least some of those viruses whose genetic material is DNA. Viral mRNA, for example, might be uniquely capable of folding back on itself to form the sort of double-stranded structure that stimulates interferon production. Once released from an infected cell the interferon binds to receptor proteins on neighbouring cell membranes, inducing these cells to make the proteins that are directly responsible for the anti-viral state. These natural "anti-viral" proteins work by interfering with the manufacture of viral proteins (and therefore new viruses) in ways that have only been partially worked out.[3] It is also possible that interferon may actually stop some viruses from getting into healthy cells in the first place.[4]

Concentrating solely on its direct anti-viral activities does not really do justice to interferon. Most people will be more familiar with its possible use against cancer than its potential to fight off the viruses. The anti-cancer possibilities stem from the still poorly understood ability of interferon to regulate various cells of the immune system, and to perhaps control cell growth in general. Natural killer cells, for example, not only *produce* their own interferon, but they are also *activated* by it. Since natural killer cells might well be important in anti-cancer defence, their activation by interferon may well be part of the explanation for interferon's anti-cancer capability; and some of the anti-viral effects of interferon might also be due to its ability to regulate the cells of the immune system, as much as its better-characterised ability to convert cells directly into an anti-viral state. At the moment there are as many questions as answers – the interferon story is certainly far from complete.

On our first exposure to any particular virus, interferon is the final major non-specific defence system to be activated. A few hours or several days after interferon production has been stimulated (the actual times involved vary quite widely for different viruses) the more famous specific defences come into play (see figure 7.2). Although the non-specific defences can be highly effective they do not display the property of our immune system that everyone has heard about: its ability to "remember" a previous infection, and so to fight it off much more effectively should it ever return. This "immune memory", or in other words our ability to become "immunised" against specific viruses, is mediated by white blood cells known as "T" and "B"-cells.[5,6]

T-cells

The first element of true specificity enters our anti-viral response when the invading viruses become bound to highly specific receptor proteins on the surface of white blood cells known as "T-lymphocytes", or just T-cells (see figure 7.3). We have

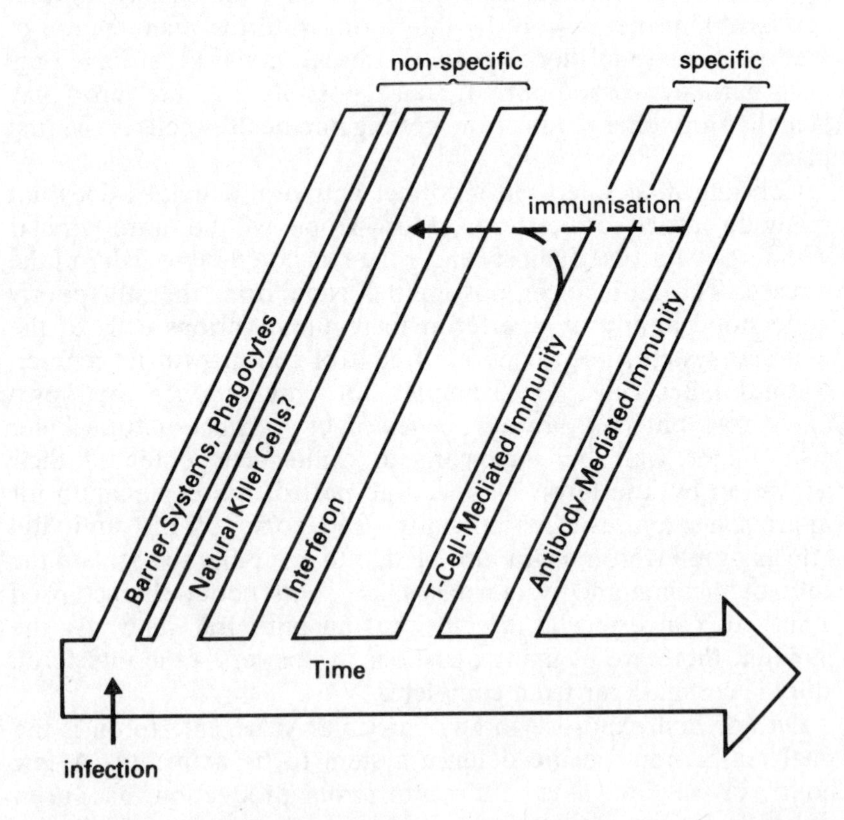

Figure 7.2 The relative timing of a typical anti-viral immune response on first exposure to the virus. On second exposure (i.e. after immunisation) the timing of the specific immune response is shifted as indicated by the dashed arrow

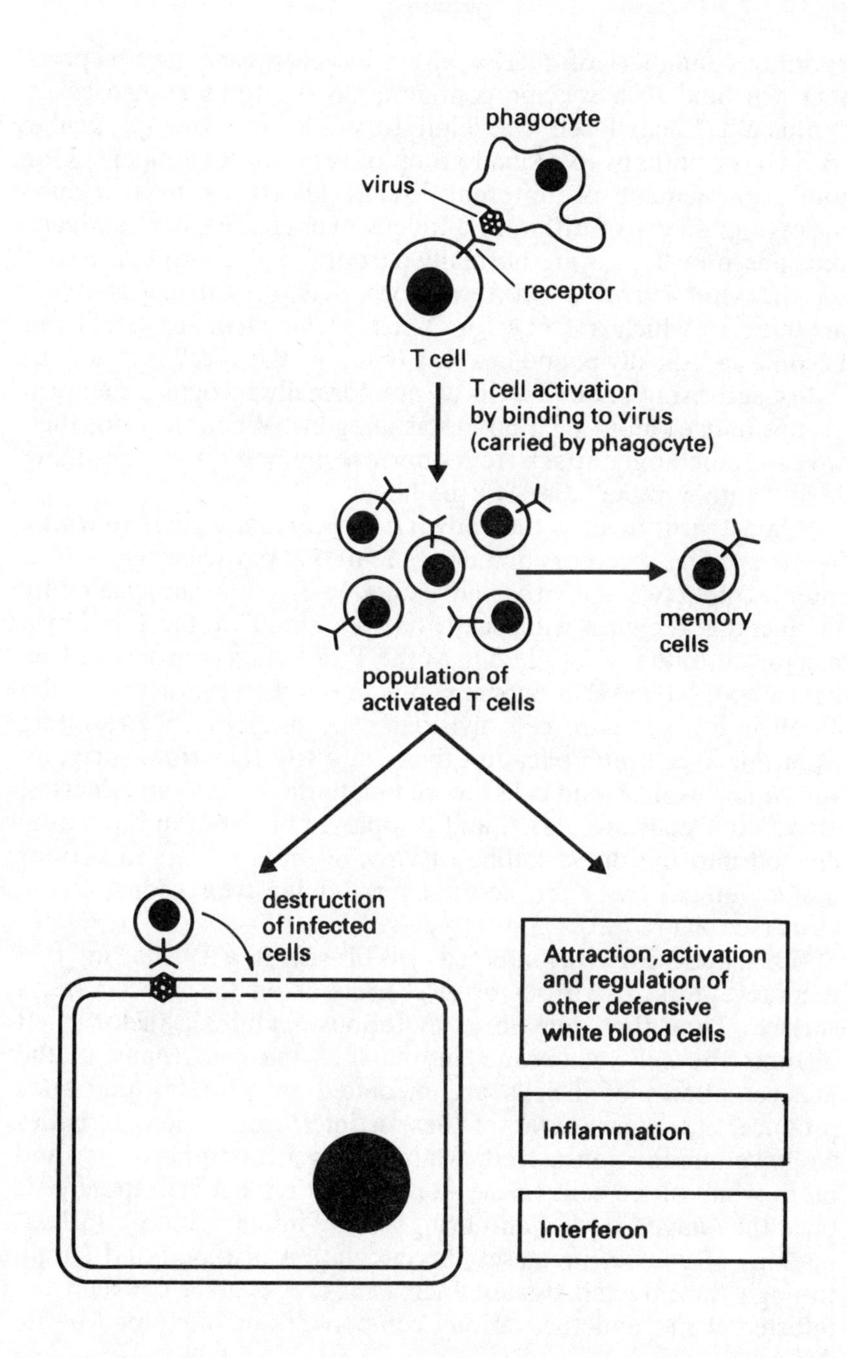

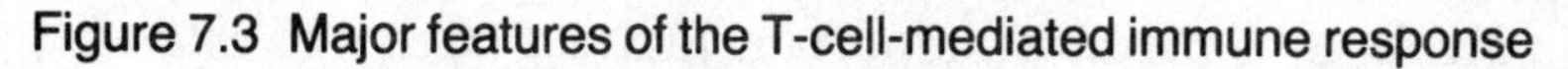
Figure 7.3 Major features of the T-cell-mediated immune response

enormous numbers of T-cells within us, each carrying receptors that can bind to a specific configuration of atoms known as an "antigen". Each T-cell can bind to either just one particular antigen, or perhaps to a small group of very similar antigens. Our total complement of different T-cells allows us to recognise massive numbers of different antigens overall. The viral antigens recognised by T-cells are normally part of the glycoproteins found on the viral surface, but remember that an antigen is really anything to which a T-cell (or B-cell, as we shall see later) can become specifically bound in a way that jolts the T-cell into action.

Not surprisingly, our T-cells do not normally recognise chemical groups native to our own bodies as antigens. When they do, then we can come under attack from our own immune system, resulting in an "autoimmune" disease (see later).

When a virus invades the body it may eventually meet up with a T-cell carrying receptors that can bind to that particular virus. The interaction between a virus and a suitable T-cell is complicated by the fact that the virus will usually be "presented" to the T-cell by a phagocyte, rather than binding to the T-cell of its own accord (see figure 7.3). However it happens, binding of the virus activates the T-cell into a series of cell divisions. The progeny cells resulting from this T-cell multiplication then work together (in association with other white blood cells) to eliminate the virus. The effects of activated T-cells are varied and complex, but they can be loosely divided into the direct killing of virus-infected cells, and various indirect effects that enhance the overall defensive response to the virus (see figure 7.3).

When attacking virus-infected cells directly, the T-cells bind (via their receptors) to viruses or viral proteins on the infected cell's surface. They then release a mysterious "cytotoxic factor" that disrupts the cell membrane and so kills the cell. Many of the indirect effects of T-cells are mediated by proteins that they produce and release. One of these is interferon, whose activities we have already considered; while others attract phagocytes and other white blood cells to the site of infection, activate these cells once they have arrived, and bring about "inflammation". Inflammation, of course, increases the circulation of blood and lymph through the infected tissues, increasing the contact between the infected tissue and the various components of our blood-borne defences.

Finally, one of the most important indirect effects of activated

T-cells is to help switch on the second stage of our specific immune defences – the "B-lymphocytes" or B-cells that make antibodies.

B-cells and antibodies

B-cells share many properties in common with T-cells. They also carry receptors that can recognise and bind to viral (and other) antigens, and binding of an antigen also activates them to multiply and act to eliminate the antigen. While T-cells eliminate the antigen by directly killing infected cells and by releasing proteins that activate our other defences, B-cells work by releasing the well-known proteins called "antibodies". Antibodies are simply proteins that can bind with great specificity to the antigens that first stimulated their manufacture, and when antibodies bind to viral antigens they can help to eliminate the virus in various ways (see figure 7.4).

Firstly, the antibodies can simply bind to the surface of free viruses (see figure 7.4c). This antibody binding might neutralise the virus simply by obscuring the sites on the virus that must be free to bind to the receptor molecules on a host cell's surface. Not all viruses, however, seem to be neutralised by antibody binding alone, so fortunately there are other possibilities available to the antibody-mediated defences. One major consequence of antibody becoming bound to a virus is that the virus becomes much more susceptible to phagocytosis. This is thanks to molecules on the phagocyte membrane that can bind to the free end of antibodies, allowing the antibodies to stick to the phagocytes and carry the viruses they are bound to with them (see figure 7.4g).

Other avenues of antibody-mediated defence involve the disruption of membranes containing either whole viruses or viral antigens that the antibodies can bind to. Infected cells often carry viral proteins in their membrane and antibodies can bind to these proteins to bring about the death of the infected cell in two main ways. Firstly, various white blood cells can bind to the free ends of the antibodies and release chemicals that disrupt the membranes and kill the cells (see figure 7.4f). Secondly, the binding of antibody to antigen activates a series of blood-borne enzymes known collectively as "complement", which can also disrupt cell membranes (see figure 7.4e). This ability to activate complement can also serve to destroy the membranes surrounding enveloped viruses (see figure 7.4d), and another important effect of comple-

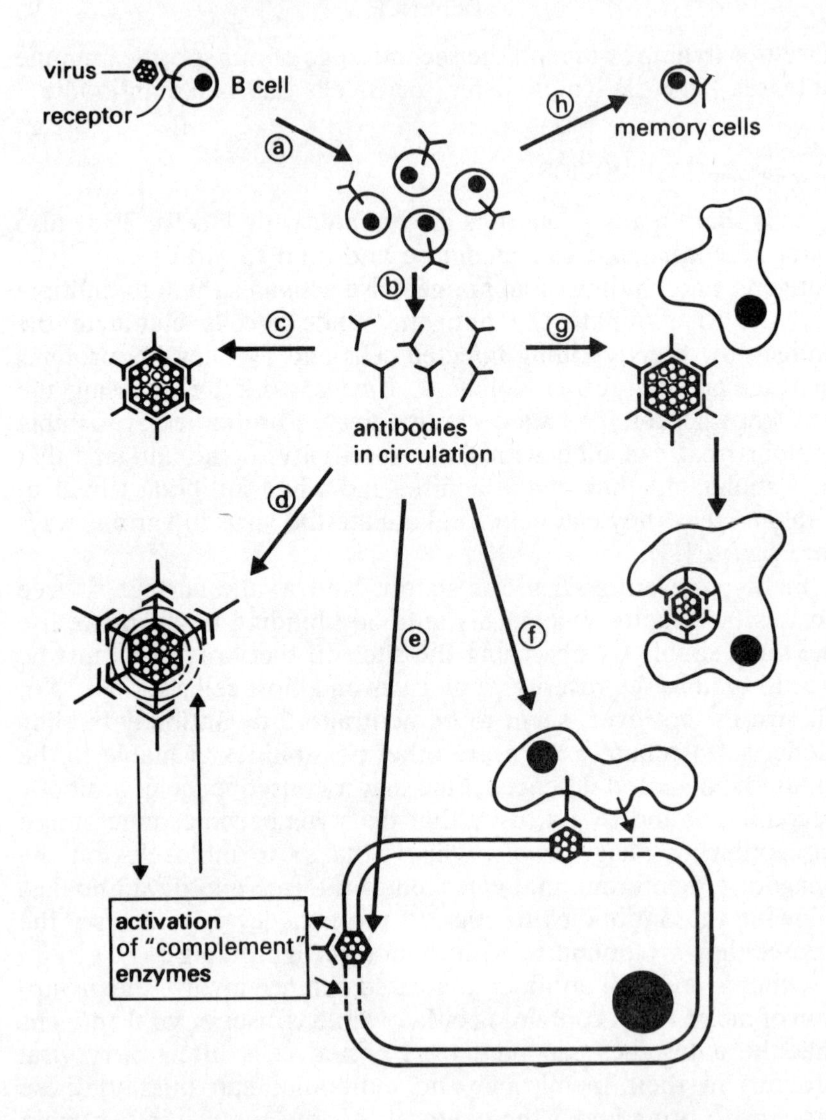

Figure 7.4 Main features of the B-cell and antibody-mediated immune response: a) B-cell activation by antigen binding; b) antibody production and release; c) virus neutralisation by antibody binding; d) neutralisation of enveloped virus by complement-mediated destruction of the viral envelope; e) destruction of infected cell by complement activation; f) destruction of infected cell by binding of cytotoxic white blood cells; g) phagocytosis of antibody-covered virus; h) production of memory cells

ment is to produce the inflammation already briefly discussed (and which will be considered again later on).

By now it must be abundantly clear to you that our resistance to viral infection is the result of a large number of extremely complex interactions, particularly between diverse types of white blood cell. The complexities of the *specific* immune system in particular are truly bewildering. Of course, I would prefer if the foregoing discussion of T- and B-cells left you with a crystal-clear understanding of specific immunity, but it is much more likely (certainly if it is all new to you) that a first reading will leave you somewhat confused. To some extent any confusion you feel might actually serve a useful purpose, impressing upon you the extreme complexity of specific immunity. To give you any truthful impression of our specific immune defences (rather than churning out the standard simplistic tale that mentions only antibodies) it really is necessary to go into the details outlined above, which still include sweeping generalisations and considerable simplifications. But having looked at some of the details of how specific immunity works, I can now much more easily tell you why one bout of infection can leave us "immune" to further attacks.

Immunisation and "memory"

The single unifying principle of the specific immune system is that each individual virus, by binding to appropriate T- and B-cell receptors, activates only those specific T- and B-cells that can effectively eliminate the virus, and as I have said, in "activating" appropriate T- and B-cells, a virus causes these cells to multiply. So the initial encounter between a viral antigen and T- and B-cell receptors that can bind to that antigen, soon leads to the production of *many more* T- and B-cells that can bind to the antigen. Now there is an important sub-population of cells shown in figure 7.3 and figure 7.4 that we have not yet considered. These are the cells labelled "memory cells", and they are the ones responsible for the phenomenon of immunisation. The memory cells are simply T- and B-cells produced during the immune response to a specific virus, and then retained in the body long after the infection that stimulated their production has been overcome. They do not really have any magical properties of "memory", but of course they carry on their surface the receptors that can recognise the viral antigens that stimulated their production. So the *next* time the same virus invades, there will be many more T- and B-cells around that can bind to the virus, via its antigens, and so initiate a second immune response. Because there will be more of these cells around, the response will be set up much more quickly than during the first infection, leading to a much more rapid and effective elimination of the virus. So the immune "memory" simply takes the form of a larger than normal number of T- and B-cells that can bind to the antigens carried by a particular virus; and the process of immunisation (either by a natural infection or vaccination) protects us against specific viruses by leaving the body supplied with plenty of specific memory cells directed against the viruses concerned.

Once you are immunised against a particular virus, then the specific immune system becomes much more significant in the early days of an infection caused by that virus. Not only will there be lots of appropriate T- and B-cells around, but a certain amount of antibody directed against the virus will always be present in your blood and secreted into the mucus lining your respiratory system and alimentary tract. The presence of this "poised" specific immune system in immunised individuals allows further bouts of the same infection to be fought off before any symptoms appear, or the effect is limited to producing a very minor illness compared to the first attack.

Clearly then, our bodies are equipped with an amazing array of powerful defences ready to fight off viral invasion. They can stop viruses from getting into us, spreading within us, infecting our cells or successfully multiplying within these cells. These defences probably save us all from disaster many times throughout our lives, as is testified to by the devastating death rate of the infamous AIDS ("acquired immune deficiency syndrome"), in which the immune system is severely impaired (see chapter 13). But of course our defences are certainly not perfect. Sometimes viruses do win the battle, at least for a short time, leading to damage and disease. The damage and diseases that the viruses can cause are the next topics to be considered.

CHAPTER 8

Damage – the virus rampant

The activities of the viruses are frequently described using the language of conflict. I have talked of viral "invasions", our anti-viral "defences" and the ability of viruses to "hijack" the molecular machinery of their host cells. Such warlike analogies make good sense when we observe the struggle from our side of the "battlefield", since the viruses clearly do invade our cells with results that can sometimes be very damaging indeed. In a very real sense we must often engage the viruses in battle if we are to survive. But from a viral point of view things look very different. It is clearly not in a viruses "best interests" to destroy the very organisms that make viral multiplication possible. Indeed, it is often said that a perfect virus (or any other parasite) would not harm its host in any way. Unfortunately, of course, many of the viruses around today are far from "perfect", since they can sometimes do great damage to individuals without threatening the survival of either mankind or the virus overall. The fact that such damage might often be the accidental result of imperfect viral evolution is hardly a comfort to us; but it might be a comfort to realise that the popular view of viruses as ruthless biological terrorists hell-bent on the destruction of living cells is completely unjustified.

The history of myxomatosis in Australia is frequently cited as evidence of the evolutionary pressures on viruses to avoid harming their hosts.[1] This well-known viral disease of the rabbit was deliberately introduced into Australia in 1950, in an attempt to control the rapidly expanding rabbit population. It promptly caused a devastating epidemic, killing almost every rabbit it infected. The Australian rabbit was clearly losing its battle with the myxomatosis virus. But the rabbit population did not die out. Instead, over a period of about 5 years the original strain of virus gave way to new and much less lethal strains. Since they killed fewer infected animals these strains were able to multiply and spread more successfully than the original strain. So evolution, driven by the natural selection of the variant viruses, was gradually tailoring the virus population to make it more compatible with the Australian rabbit. Of course the rabbit population would also be evolving to some extent, through the survival of those animals that were less susceptible to the effects of the virus.

It is of no advantage to a virus then, to inflict *gratuitous* damage on its host. This comforting thought, coupled with your knowledge of the extremely effective defences we have available to repulse viral invasions, should make it easy for you to accept that *most viral infections do us no real damage at all*. For obvious reasons we tend to be preoccupied with the diseases that viruses can cause, but disease is actually the exception rather than the rule. For every time throughout our lives that we fall ill due to viral infection, there are many more occasions on which viruses set up infections that we never notice. Such "inapparent" infections are doubtless currently in progress in many of the apparently healthy people you encounter each day. They can last for a few days, or a lifetime; and it is feasible that every one of us, all of the time, is harbouring some quiet and unnoticed viral infection.

This does not mean that there is one broad class of viruses that never produce disease, and another class whose members always make us ill. Instead, with most types of virus the course of an infection can often vary between the two extremes of harmlessness and serious illness. Poliovirus is an excellent example. Prior to vaccination programmes it caused great disability and suffering throughout the Western world, and it continues to be one of the most damaging viruses of all in many parts of the Third World. But despite its gruesome record and terrible reputation, poliovirus infection is normally completely harmless. Usually the infection remains restricted to the throat and intestines, causing little harm until it is eventually eliminated by the body's defences. Only very occasionally does the virus gain access to the nervous system and destroy the crucial nerve cells whose death produces the severe paralytic disease we call "polio".

Inapparent infections, caused by potentially dangerous viruses such as poliovirus, might be vitally important to the spread of a disease while being of no importance to the fortunate individuals infected. A poliovirus infection that remains harmlessly restricted to one person's throat and intestines, allows the virus to multiply and then return to the environment via faeces. Viruses produced by one bout of inapparent infection might then go on to cripple the next victim. But inapparent infections can also bring their benefits, for they may leave the infected individual immunised against further infections caused by the same (or similar and perhaps more dangerous) viruses. So those inapparent viral infections that are with us much of the time may be quietly spreading both the virus, and immunity to the virus, throughout the population. Although unnoticed, they may be crucially important to the epidemiology of disease overall.

Cell death, cell change

Having spent some time emphasising that viruses certainly do not always harm us, and indeed may often be subject to strong evolutionary pressure to develop into progressively more harmless forms, I must now return to the many damaging things that they undoubtedly can do. A quick glance down a list of human diseases caused by the viruses (figure 3.5 p. 42 for example) confirms that the viruses can sometimes do us a great deal of harm. The obvious questions prompted by such a list are firstly, how do viruses cause these diseases, and secondly, why are the diseases they cause so varied? We have already considered one factor that allows the viruses to cause different diseases, namely their variable ability to infect and spread to specific types of cell, and we will be looking at other factors later. But what about the first question – how do the viruses actually cause disease? Indeed, why should they cause disease at all, considering the evolutionary pressure encouraging them not to harm their hosts?

The simple answer that many viruses are *imperfectly evolved*, doing some damage to their hosts but not enough to make either virus or host extinct, is probably part of the explanation. But it is also important to realise that the damage caused by viruses is rarely completely gratuitous. Often it is an inevitable result of the processes that allow viruses to get inside, multiply quickly within, and then escape from the infected cell. It would actually be rather difficult to design the so-called "perfect" virus that would cause absolutely no harm to its host while rapidly spreading throughout the host population. The accounts of evolution are recorded simply in terms of reproductive success, allowing some degree of damage to a virus's host to be tolerated provided the returns in terms of increased viral reproduction are sufficiently large. So the life-cycle of many viruses is probably a compromise between the need to multiply and spread rapidly, and the evolutionary pressure not to do too much harm in the process.

All of the harm that viruses do to us must ultimately be explicable in terms of the effects of viral infection on individual cells. There are two main damaging things that viruses can do to the cells they infect – *they can kill them or they can change them.* Both of these effects can be brought about either directly by the viruses themselves, or indirectly as a result of the body's immune response to the infection.

You are no doubt becoming well used to the apologetic statements like "the precise details of this aspect of virology remain to be uncovered", scattered throughout this book. Such statements are particularly applicable to what is known as the "molecular pathogenesis" of viral infection: in other words the precise chemical details of what is going on when viruses kill or change cells. We certainly know of many damaging *effects* that viruses have on infected cells, but it is much easier to observe and describe such effects than to actually determine how they come about.[2,3]

Concentrating for the moment on the direct effects that viruses can themselves have on cells (leaving indirect damage due to the immune system until later), you will find a summary of the most important effects in figure 8.1. Glancing quickly at some of the things that viruses can do to cells should make it clear that they have plenty of scope to cause the cell death and disintegration associated with many infections. Infections that kill cells are often accompanied by the complete shut-down of the manufacture of host cell proteins, RNA and DNA. By closing down these central cellular production lines a virus can eventually bring all normal cellular activity to a halt. Crucial structures such as cell membranes will begin to disintegrate due to lack of maintenance and the cell will die. Often, cell death might be accelerated by the release of powerful degradative enzymes from lysosomes. As was briefly mentioned earlier, these are small membrane-bound structures that normally process materials brought into the cell from outside.

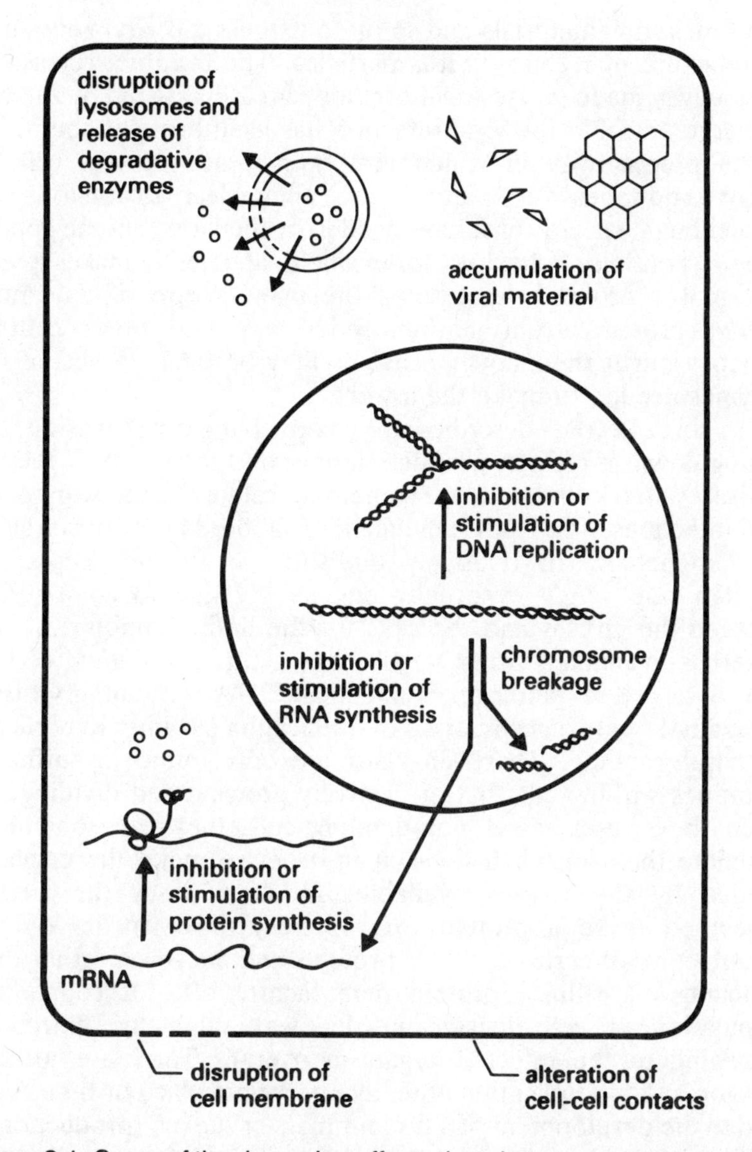

Figure 8.1 Some of the damaging effects that viruses can have on cells

The processing enzymes, once released into the cell at large by disintegration of the lysosome's membrane, might wreak havoc on the delicate structure of the cell.

Prior to cell death, however it is brought about, the invading viruses will be able to exploit the pre-existing cellular enzymes, ribosomes, raw materials and so on to express the viral genes and manufacture many more virus particles. The eventual release of these newly made viruses might then be greatly assisted by the cell disintegration that the viral infection has itself brought about.

The precise way in which viruses sabotage normal cellular activities doubtless varies and is often still unclear. In at least some cases though, specific proteins encoded by the viral genome appear to be responsible. Poliovirus, for example, apparently makes a viral protein (or proteins) that stops the manufacture of any more *cellular* proteins from getting under way. All other cellular activities might then slowly cease as they become starved of the enzymes needed to make them work.

The sort of events described above, involving viral invasion, the closing down of cellular production lines and the eventual release of viruses after cell death, are sometimes related as the story of all viral infections. The truth is much more subtle. Many viruses leave cellular metabolism relatively undisturbed. In such cases the infected cells *might* eventually die, simply due to competition between the viruses and the cells for the limited amount of raw materials available. Other viruses can actually *stimulate* certain aspects of cell metabolism. Some small DNA-containing viruses, for example, rely on cellular DNA-replicating enzymes to replicate the viral genome. These enzymes are only found in sufficient quantities within cells that are actively growing and dividing. So when the viruses infect non-dividing cells they can sometimes stimulate them into cell division in order to make the enzymes needed by the viruses available. In these cases the viruses concerned make a protein (or proteins) that *switches on* the manufacture of certain cellular proteins and enzymes, rather than switching all cellular protein manufacture off. Of course the stimulation of cell division in this way might be extremely damaging to the infected organism overall. The onset of cell division and multiplication at an inappropriate place or time could lead to the development of a tumour mass or the overproduction of hormones or other specialised products of the infected cells.

Some viruses are known to alter the expression of particular cellular genes without damaging the infected cells in any other way. One very interesting example is the recent discovery of a virus that can cause mouse pituitary glands to underproduce growth hormone.[4,5] Infected mice are clearly smaller than normal, but are otherwise unharmed. Even right down at the level of individual cells the virus seems to do no damage other than restrict the production of the hormone. The discovery of viruses such as this is prompting many scientists to investigate in detail the subtle roles that viruses might play in many poorly understood diseases. Clearly, we are now talking about the *changes* that viruses can bring about in infected cells, rather than their more dramatic and better-publicised ability to cause cell death. One of the most damaging changes in cell activity that viruses can induce is of course the initiation of the uncontrolled cell division that causes cancer. This is a subject of such importance and recent interest that it has been given its own chapter (chapter 10).

The changes that viruses can bring about within cells are by no means restricted to the switching on or off of cellular genes. Many viruses (herpes and measles viruses for example) can actually cause cell chromosomes to break in two. This might be a secondary effect of the shut-down of cellular protein synthesis, which will bring to a halt the manufacture of the enzymes that normally repair such damage, but perhaps not. Whatever causes it, the consequences of chromosome breakage are not clear, but obviously such dramatic damage to the cell's genetic material might have a devastating effect on cellular activity overall.

The outer cell membranes that determine the interactions between neighbouring cells are another site at which viruses can interfere with the activities of a cell without necessarily killing it. Changes in the cell membrane, caused by either the introduction of viral proteins or an alteration in the complement of cellular proteins, are found in many viral infections. An example of the possible results is the fusion of herpes or measles virus-infected cells into massive complexes containing many cell nuclei. Of course the cell membrane is also the site at which the many elements of the immune system interact with an infected cell. The interaction of both infected cells and free viruses with our immune defences brings us to a whole new range of ways in which viruses can bring about damage and disease. For although it clearly saves us from many of the consequences of viral attack, the immune system may sometimes cause damage far worse than the damage a virus would itself have caused if left unchallenged.

Immune damage

The ability of our immune defences to cause us harm comes as quite a surprise to many people. It contradicts the popular view of the immune system as a triumph of evolution perfectly "designed" to combat infection. In fact, while our immune defences are perhaps a "marvel" of evolution they are by no means ideal. During the immune response powerful forces are activated to kill infected cells and eliminate viruses. Unfortunately these forces cannot readily discriminate between life-threatening infections that absolutely *must* be overcome, and relatively harmless infections that might be better left alone.

We saw earlier that infected cells can be killed in a variety of ways when antibodies or activated T-cells bind to viral materials on the cell surface. Much of the liver damage found in serum hepatitis seems to be caused when the immune system kills cells in this way, and cell killing by the immune system probably also accounts for much of thc damage done by herpesviruses and poxviruses. Of course it is usually quite difficult to estimate how much damage the immune system causes compared to the direct effects of an infection; or whether or not an illness would be milder if certain aspects of the immune response were avoided. But experiments on mice have confirmed that in at least some cases the immune system provides "help" that an organism could well do without. A virus known as LCM virus can infect the mouse nervous system and cause severe meningitis and eventually death, but the infection itself seems to be completely harmless. Indeed, if the mouse immune system is artificially suppressed then illness is prevented and the infected animal survives.[6] So here we have an example of the immune system killing an animal, rather than saving it; demolishing the notion of immunity as a perfect defence system crafted by millennia of entirely beneficent evolution.

Apart from directly killing cells infected with relatively harmless viruses, the immune system can also cause considerable indirect damage by activating an inflammatory response. We saw earlier that inflammation is caused when T-cells release certain proteins and when antibody-antigen complexes activate the enzymes of the complement system. Other activators of inflammation are released from damaged cells, such as those destroyed during many virus infections. The term "inflammation" actually describes a wide range of effects that are not fully understood, but the results are frequently obvious: the circulation of blood and lymph through inflamed tissue is greatly increased, bringing large numbers of the white blood cells responsible for immunity to the place where they are most needed. The increase in circulation causes the redness and swelling common at the site of infection or injury, and the swelling can put pressure on nearby nerves to cause pain. Inflammation is a "good" thing, in that it assists the immune response to an infection, but if the infection is harmless or the inflammation unduly vigorous it can cause a lot of unnecessary damage and pain.

Again, there are examples from the laboratory of the damaging effects of inflammation. A virus known as Borna virus can set up a long-term infection in the brains of rats, leading to frenzy and eventually blindness. But if rats are made "tolerant" to the virus (meaning that they won't recognise it as foreign and so will not set up an immune response against it), then the illness can be avoided. In this case the disease is apparently a direct result of the inflammation that occurs in normal rats soon after infection.[7]

The damaging effects of inflammation need not be restricted to the site at which a viral infection is actually taking place. In some infections large complexes of antibodies bound to viral antigens (known as "immune complexes") can travel through the circulation until they become stuck in the small capillaries found throughout the body, particularly in the kidneys. Once lodged in the capillaries the immune complexes can produce inflammation far from where it is really required (by activating complement enzymes, for example). Tissue damage produced by such immune complexes is found in many long-term or "persistent" viral infections. It probably explains much of the damage to the kidneys and circulatory system that can accompany serum hepatitis.[8]

If you were to suffer from any of the above examples of immune damage then you might reasonably conclude that your own immune system had "turned against" you. But in all the cases looked at so far the damage is an indirect effect of an immune response that remains directed against the invading viruses themselves. The final and most fascinating mechanism of virus-induced immune damage involves a direct immune assault against the host's own tissues; for sometimes viruses can bring about a state of "autoimmunity", in which the cells of the immune system mistakenly recognise some normal host proteins (or other materials) as foreign antigens. This leads on to a full-blown immune attack against some of the host's own healthy tissues and cells, which can obviously do considerable damage and cause serious disease. Anyone unfortunate enough to develop autoimmunity really can claim that their immune system has turned against them.

The role of virus-induced autoimmunity in specific human diseases is still somewhat confused and often speculative, but it has been implicated in such serious conditions as multiple sclerosis, rheumatoid arthritis and diabetes (see chapter 13). Quite a few plausible mechanisms by which viruses might switch on auto-immunity have been suggested. One of the most obvious possi-bilities involves viruses that carry antigens very similar to some proteins found on or within their host cells. There is some evidence that evolution can tailor viral proteins to look like host cell proteins when "examined" by the immune system. Ultimately, of course, this might allow a virus to avoid our immune defences altogether, since we do not generally recognise our own proteins as foreign. But if the match between viral and host proteins is close but imperfect, then an immune response set up against the viral antigens might also be effective against the similar parts of the proteins of healthy cells (see figure 8.2a). Viral antigens that "look" very like normal cell proteins to the immune system certainly do exist. Antibodies have been found, for example, that bind not only to antigens carried by measles and herpes simplex viruses, but also to a filamentous protein that is a normal constituent of human cells.[9]

Another possible way for a virus to induce autoimmunity is by altering the expression of host cell genes, so that cell proteins not normally exposed to the immune system become available on the cell surface. These unusually sited normal proteins would then be regarded as foreign (since the immune system cells would not be used to "seeing" them) and thus would be subjected to immune attack (see figure 8.2b). Alternatively, the destruction of cells by viral infection might release internal cell proteins normally hidden from the immune system (see figure 8.2c), or complexes formed between viral and cell proteins might make the cell proteins "look different" and therefore foreign (see figure 8.2d). The immune response set up against such complexes might then also attack, albeit less effectively, the host cell proteins in their un-complexed state. Then there is always the possibility that by infecting T- and B-cells directly, a virus might alter these cells' activities and make them attack the host cell proteins which they would normally leave alone (see figure 8.2e).

These are only some of the ideas put forward to explain the

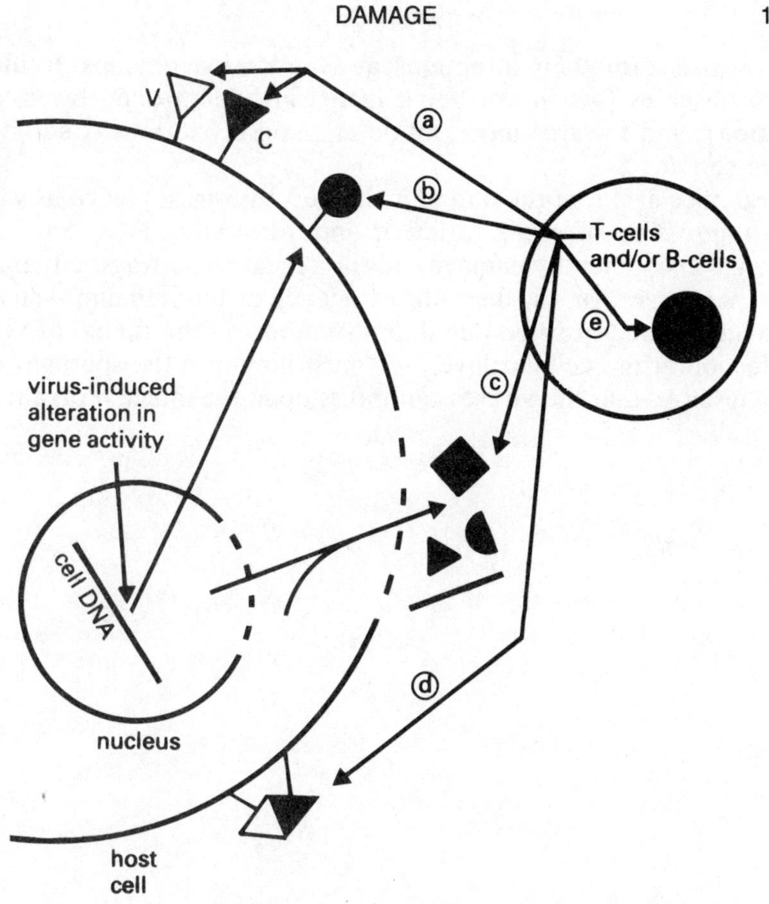

Figure 8.2 Ways in which viral infection might induce autoimmunity by making the immune system wrongly identify host cell materials as foreign antigens: a) Attack on cell protein C that "looks" like viral protein V; b) attack on abnormally produced and sited host cell protein; c) attack on materials released from dying cell; d) attack on complex formed between viral and host cell proteins; e) viral infection of T- or B-cells, causing cells to behave abnormally and attack the cells of the host

autoimmunity that accompanies or follows on from many different viral infections. Since the subject is still an area of considerable mystery and confusion, there is little point in going further into more precise and possibly mistaken details. The involvement of viruses in autoimmunity will come under increasing scrutiny as virologists turn their attentions away from the obvious "acute" viral diseases (which are being increasingly controlled by vaccination), and towards more subtle and still mysterious conditions (see chapter 13).

So once again, at the end of a chapter, the viruses leave us with an impression of great variation and versatility. They can kill, change and generally damage cells in a great many ways; either all by themselves, or via the indirect agency of the immune system. Having looked in some detail at variations on the theme of viral infection at the cellular level, we must now turn the spotlight on the diseases that the viruses can inflict upon the infected organism overall.

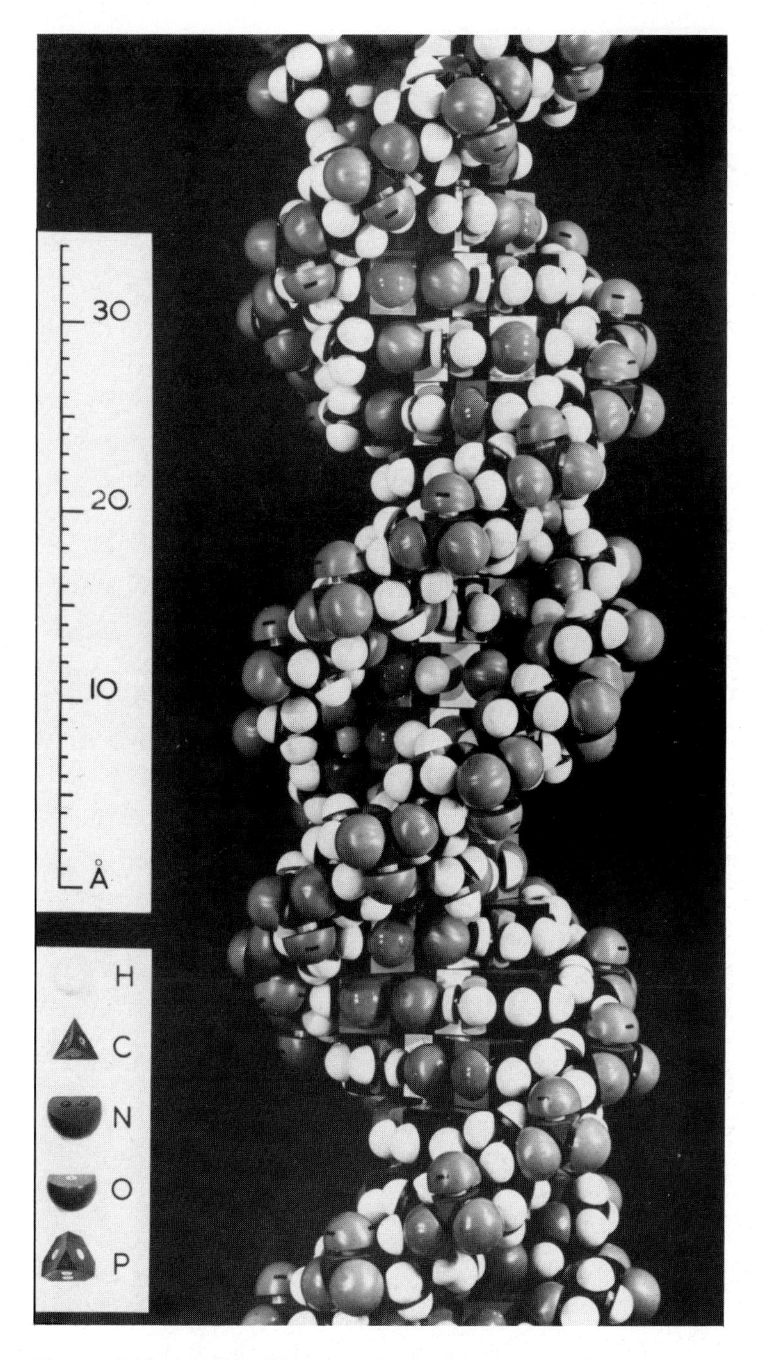

Plate 2.1 Model of the DNA double-helix, with plastic shapes representing individual atoms. One Angstrom (Å) = 10^{-10} metres
(By courtesy of the Biophysics Department, King's College, London)

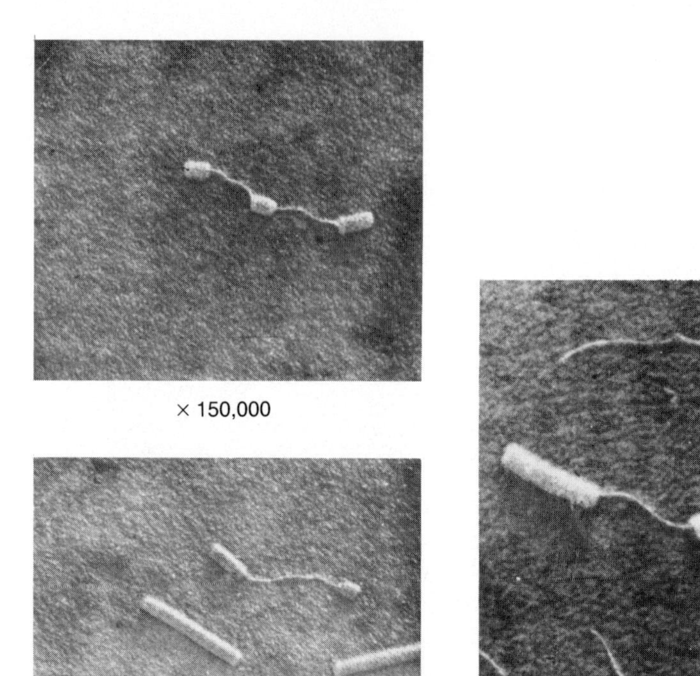

× 150,000

× 150,000

× 300,000

Plate 3.1 Tobacco Mosaic Virus particles with sections of their protein coats stripped off to reveal the central threads of single-stranded RNA
(By courtesy of M. K. Corbett, University of Maryland. From Virology *(1964) vol. 22, p. 539 © Academic Press)*

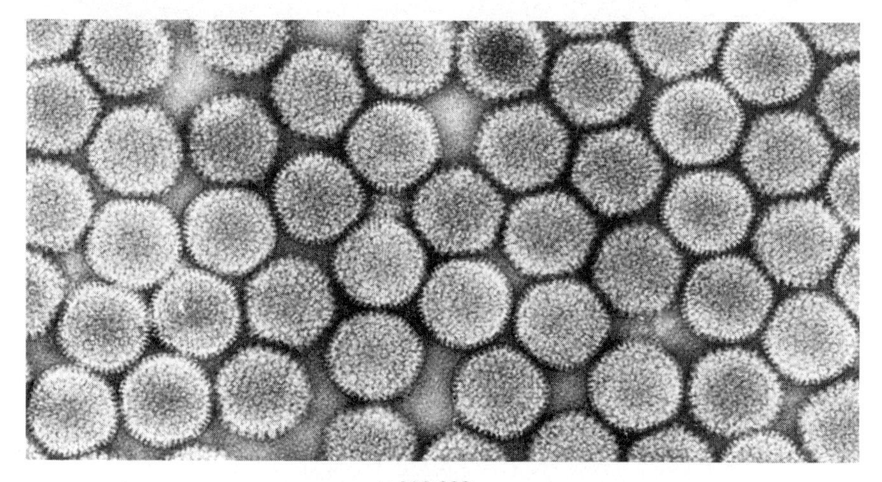

× 200,000

Plate 3.2 A cluster of adenovirus particles
(By courtesy of R. W. Horne, University of East Anglia)

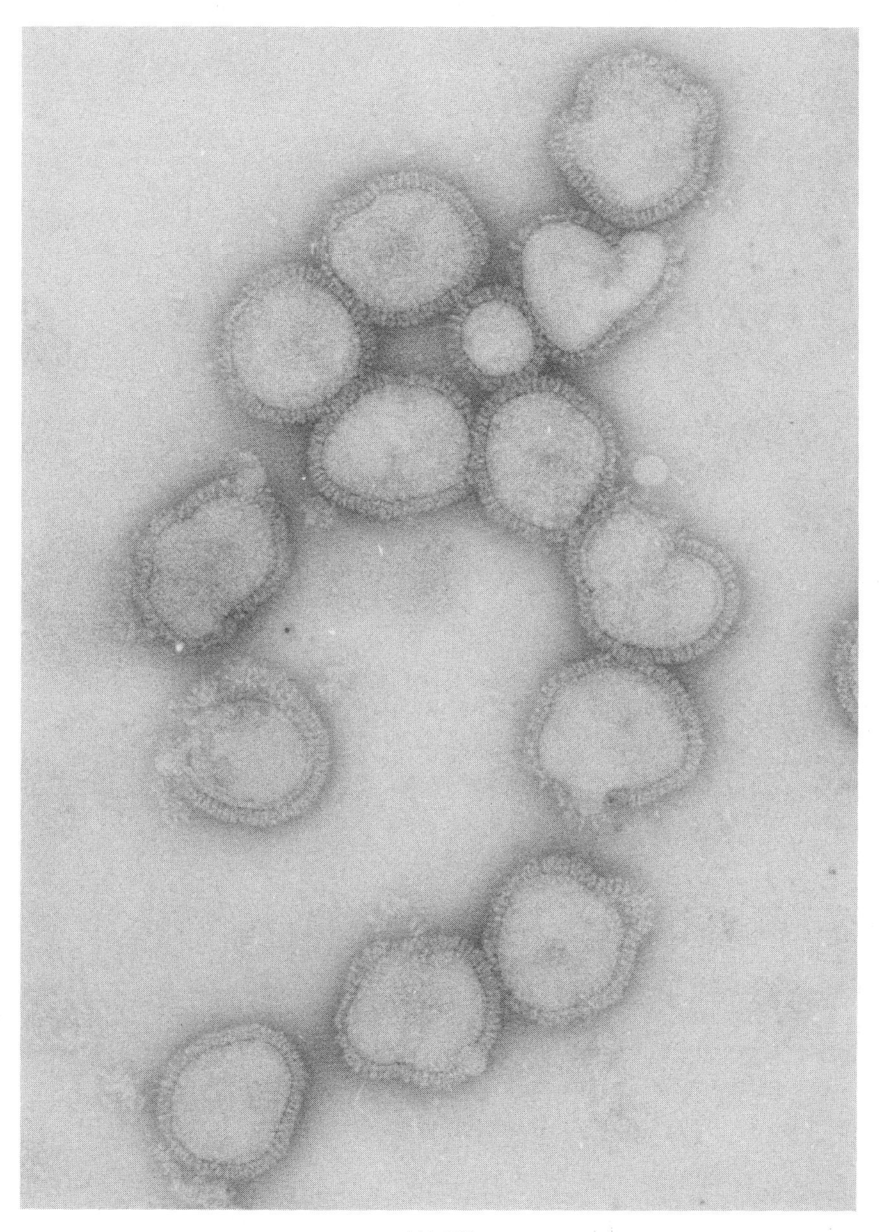

×140,000

Plate 3.3 Influenza virus particles
(By courtesy of I. T. Schulze, St. Louis University Medical Center. From Advances in Virus Research *(1973) vol. 18 pp. 1–55 © Academic Press)*

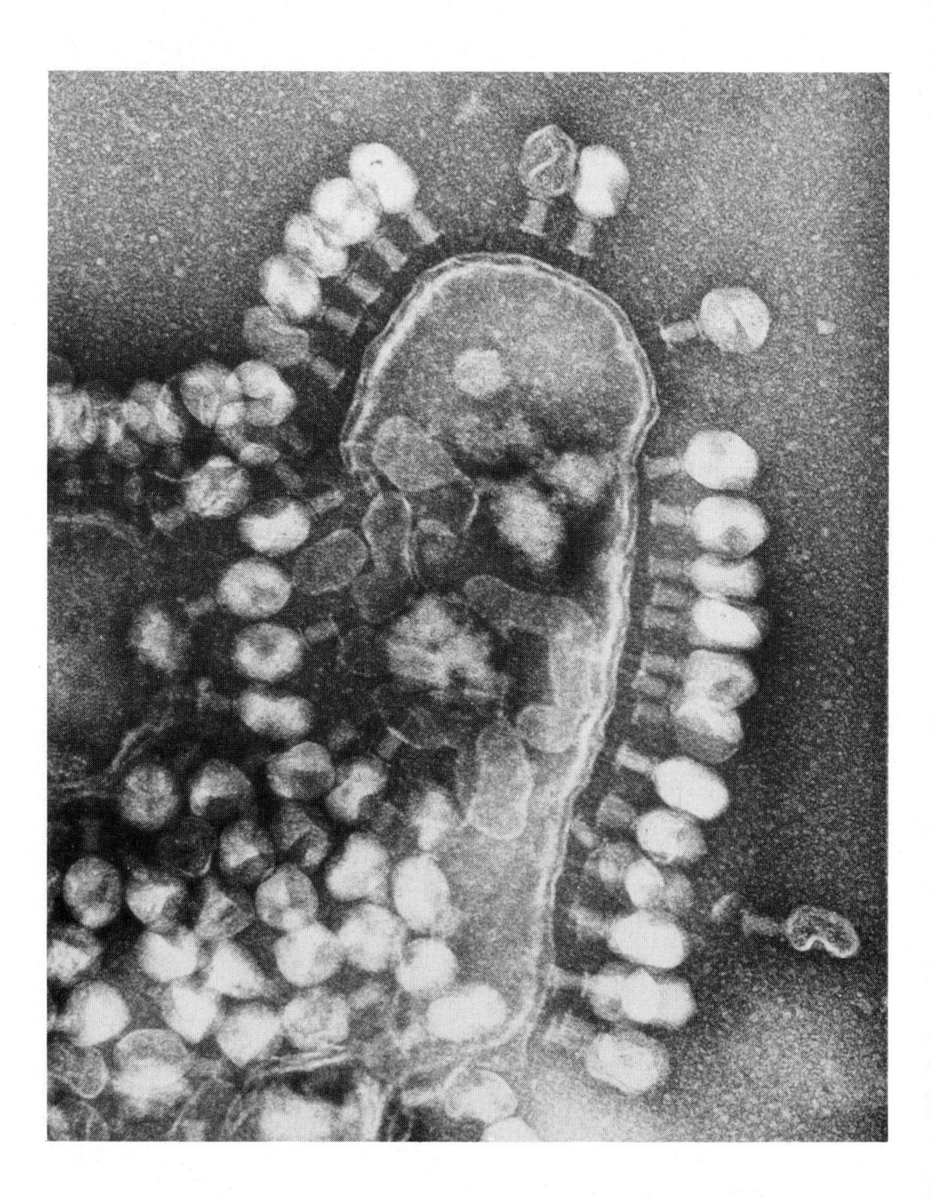

Plate 3.4 T4 Bacteriophage particles attached to an infected bacterial cell
(By courtesy of T. F. Anderson, Fox Chase Cancer Centre and L. D. Simon. From Virology *(67) vol. 32, Fig 6 p. 285 © Academic Press)*

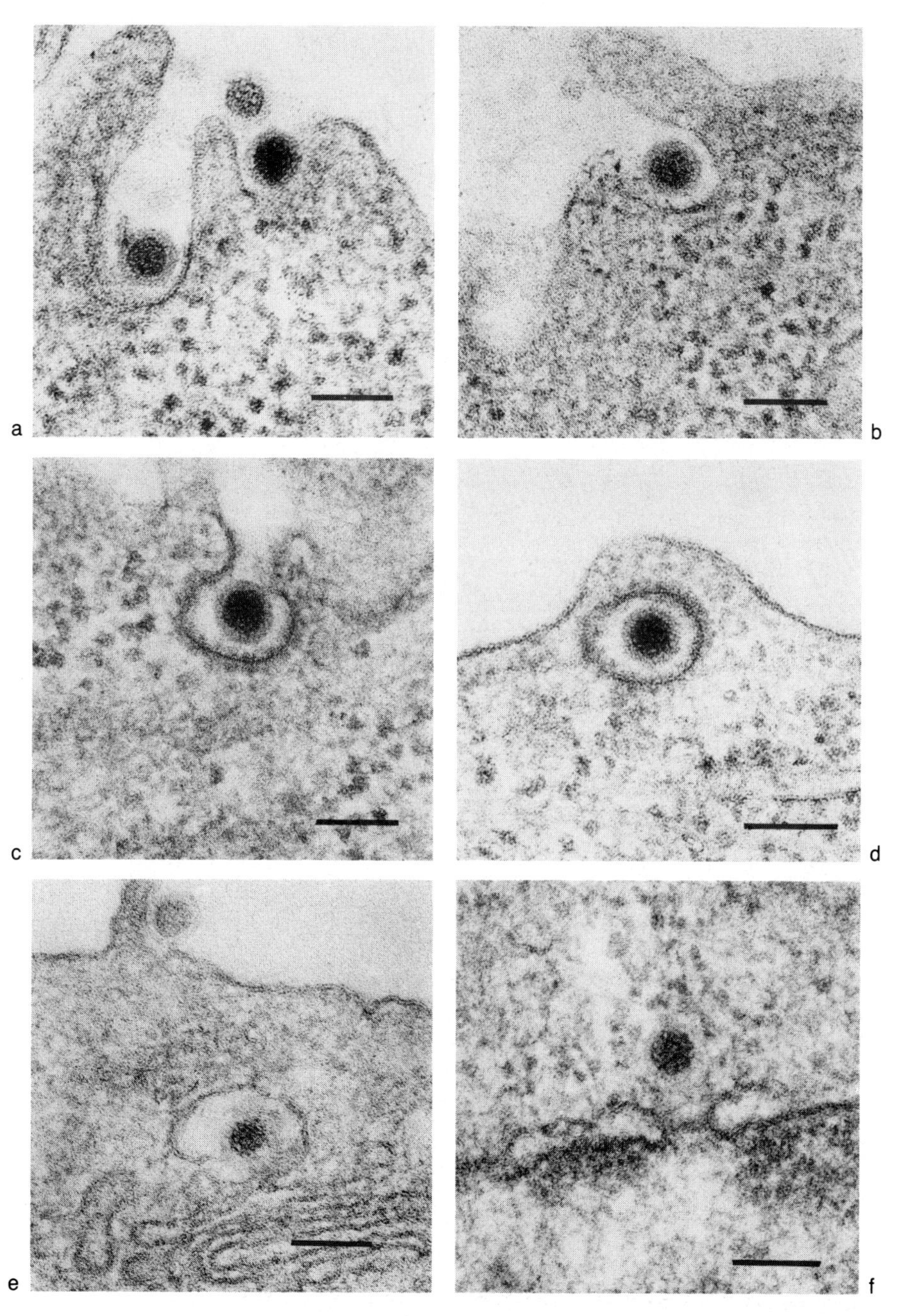

Bar marks = 100 nm = 10^{-7} metres

Plate 4.1 Entry of adenovirus by endocytosis. a) Virus at cell surface. b–e) Formation and passage into cell cytoplasm of virus-containing vesicle. f) Free virus inside cell
(By courtesy of S. Patterson, Clinical Research Centre, Harrow, Middlesex)

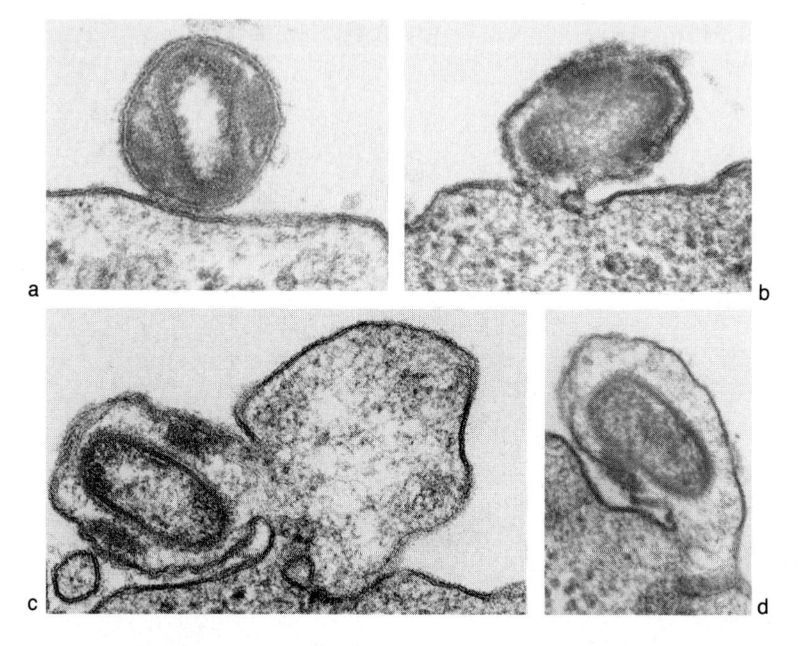

Plate 4.2 Entry of a pox virus by membrane fusion. a) Virus at cell membrane b) Fusion of viral and cell membranes. c and d) Virus core becoming surrounded by cell cytoplasm.
(By courtesy of J. A. Armstrong, Royal Perth Hospital, Australia)

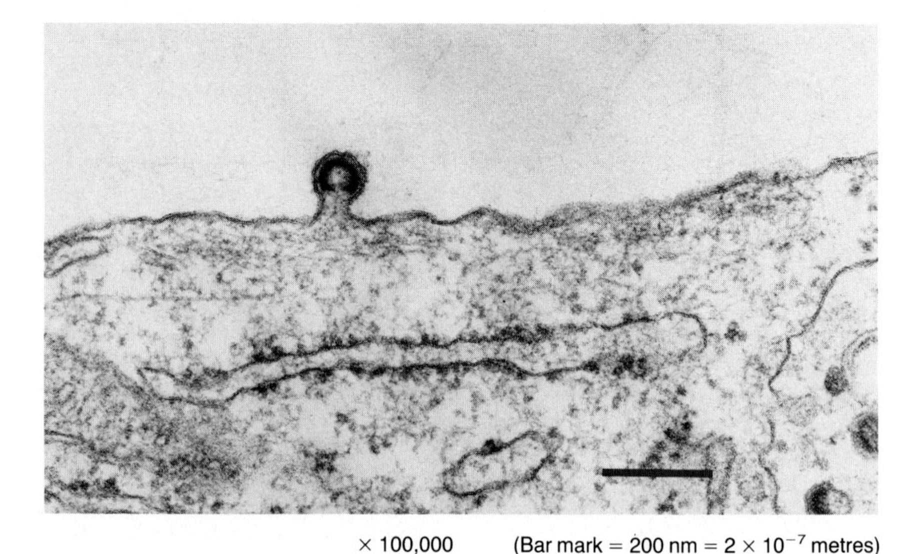

× 100,000 (Bar mark = 200 nm = 2×10^{-7} metres)

Plate 4.3 A retrovirus escaping from an infected cell (and acquiring its outer membrane) by 'budding' from the host cell membrane
(By courtesy of S. Patterson, Clinical Research Centre, Harrow, Middlesex)

Plate 6.1 The dispersal of 20,000 or so droplets following a violent sneeze
(from 'The Pathogenesis of Infectious Disease' by C. A. Mims, from Aerobiology *(F. R. Moulton) ed., fig 5, p. 118, by M. W. Jennison © 1942, by AAAS)*

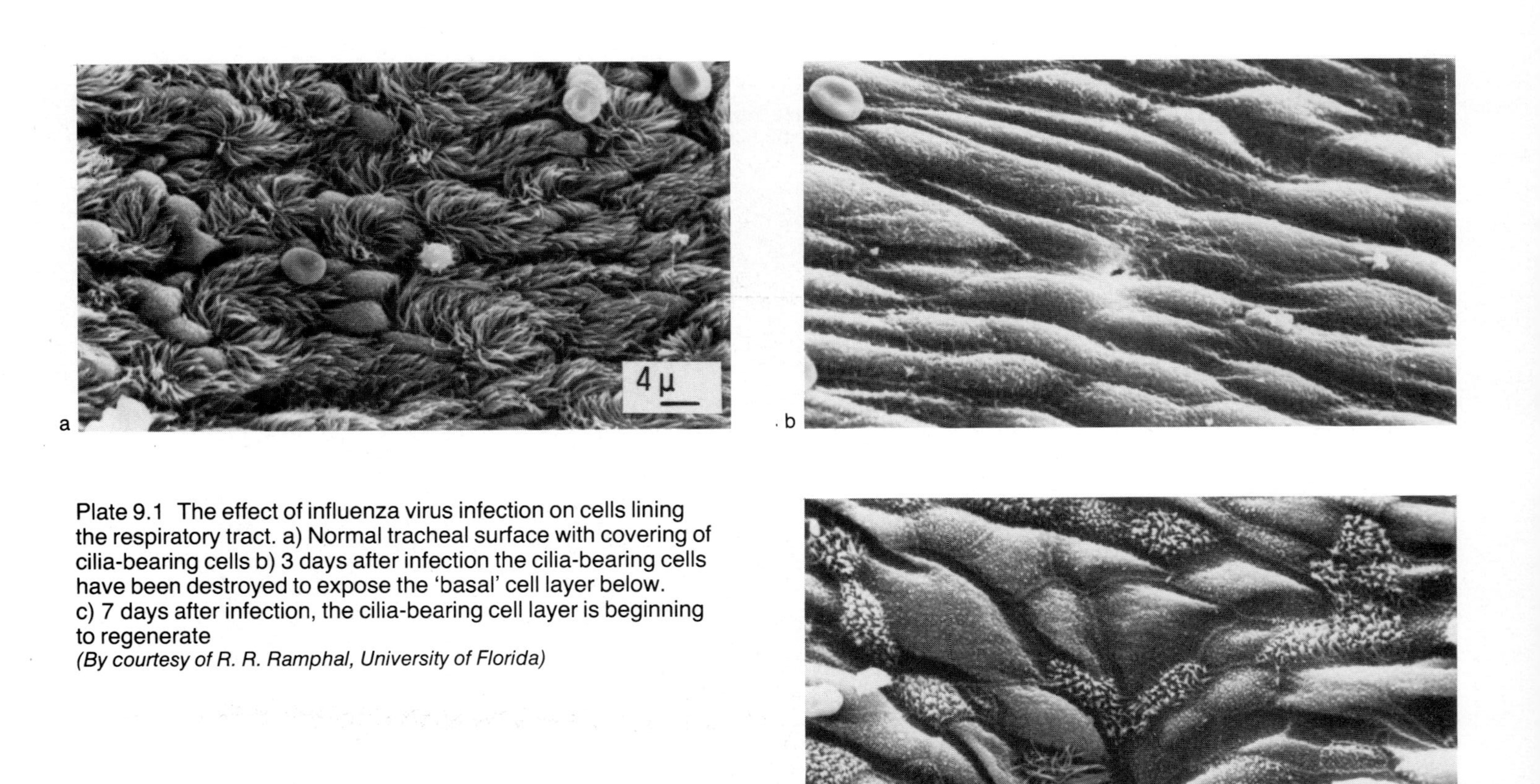

Plate 9.1 The effect of influenza virus infection on cells lining the respiratory tract. a) Normal tracheal surface with covering of cilia-bearing cells b) 3 days after infection the cilia-bearing cells have been destroyed to expose the 'basal' cell layer below. c) 7 days after infection, the cilia-bearing cell layer is beginning to regenerate
(By courtesy of R. R. Ramphal, University of Florida)

$1\mu = 10^{-6}$ metres

CHAPTER 9

Disease – an infinity of courses

If I were to say that the viruses can do almost anything to us it would only be a slight exaggeration. Viral infections can kill us quickly or pass harmlessly and unnoticed. A bout of acute illness might be followed by complete recovery, or may develop into a persistent chronic illness that might eventually result in death. Alternatively, chronic illness might be produced from the outset, or an initially inapparent infection may surface as either an acute or chronic illness many years later. The symptoms accompanying all these possibilities will vary widely, depending on which organs are infected, which viruses are involved and a host of other often poorly understood considerations.

There are actually so many alternative patterns of viral disease that it would be impossible to incorporate them all into one meaningful diagram. What we can do, however, is to identify the four broad and imperfect categories shown in figure 9.1. These represent clearly distinguishable extremes of the almost infinitely variable phenomenon of viral infection. Having defined these categories, we should be able either to fit any infection into one category alone, or else describe its sequential development from one category to another.

To put some flesh on the four bare bones of figure 9.1 we must take a look at exactly what is happening when the viruses make us ill. Now I certainly do not want to present you with a comprehensive survey of viral diseases, while generalising everything into a catalogue of effects with no mention of specific diseases would be equally unhelpful. Instead, having already considered inapparent infections, this chapter will use specific examples to illustrate the

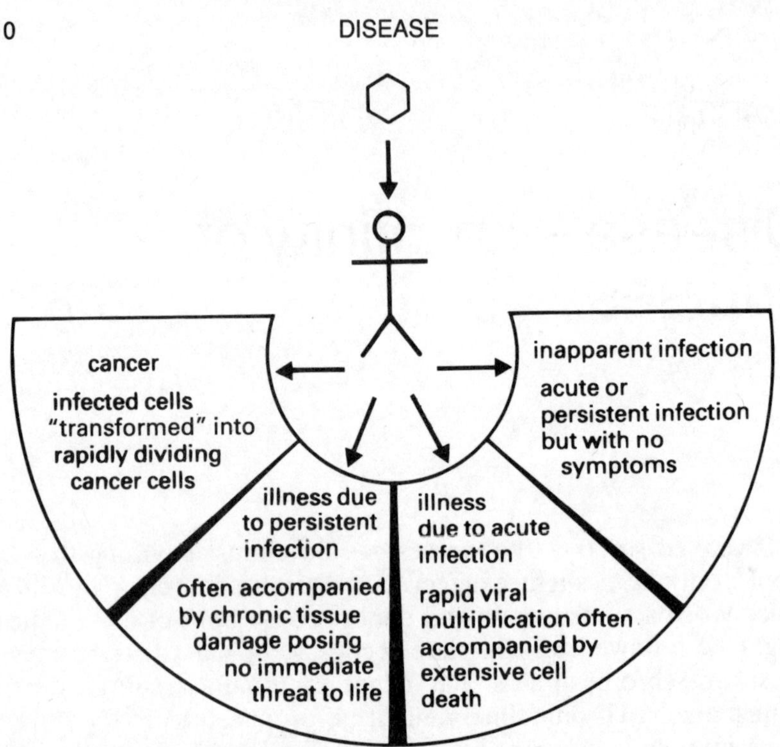

Figure 9.1 The major alternative overall consequences of viral infection

principles of "acute" and "persistent" infections in turn. Chapter 10 will deal with the involvement of viruses in cancer, while in chapter 13 I will look at some currently mysterious diseases that might well be viral in origin.

Acute infection – influenza

When virologists talk of "acute" infections they mean those in which the virus enters and rapidly multiplies within some region of the body, followed by the equally rapid elimination of the infection by our defences or the swamping of these defences leading to death. So whatever the final outcome is, a straightforward acute infection is a *brief* encounter between a virus and its host. Of course besides sometimes causing death, such brief encounters can also produce a wide range of "acute" viral diseases. Acute infection represents the virus–host interaction at its simplest level – a lively conflict followed by a swift conclusion. If the invading virus is defeated then it may disappear from the body altogether, leaving nothing behind other than an expanded population of immune memory cells ready to counter the virus more quickly and effectively should it ever return.

There are, of course, a great many viruses that normally follow the pathway of acute infection, causing illnesses such as measles, mumps, gastroenteritis, meningitis, rabies and the common cold. But one of the best examples with which to illustrate the characteristics of acute infection is influenza – a dramatic, brief and debilitating illness which reduces most of us to feverish helplessness at least once in our lives.[1]

When influenza strikes there is little we can do but lie and wait for it to be defeated. Thankfully it normally *is* defeated, but its widespread and repeated occurrence transforms a relatively low death rate into a large number of deaths overall. The great influenza pandemic of 1918–1919 killed about 20 million people in only a few months – one of the most devastating attacks of infectious disease ever experienced by mankind. And pandemics of influenza continue to sweep the globe every 15–20 years, interspersed with more localised epidemics that arrive every year or so. Several viruses vie for the title of the biggest killer of them all and ownership of the title is unlikely to ever be firmly established; but influenza virus (in association with the secondary bacterial infections that it leaves the body vulnerable to) is certainly one of the chief contenders.

The virus is inhaled into the body and quickly sets up an acute multiplicative infection of the upper respiratory tract. It can enter and multiply in the cells lining the nose, throat and windpipe, with the windpipe as probably the major site. Within only 2 or 3 days the infection can have spread over a large area and the characteristic symptoms of "the flu" will then begin: chest and throat pains accompanied by a dry cough, fever and possibly even delirium, headaches, muscle pain, shivering and a general feeling of weakness and lethargy. Our problem is to explain how the multiplication of a virus in the cells lining the respiratory tract can do all this (see figure 9.2).

Let's start by looking at the damage the virus does to the cells it actually invades. The cells lining the respiratory system have short fibres called "cilia" sticking out from their surface. The rhythmical beating of these cilia spreads the protective mucus over the airways of the lungs and throat, moving it upwards from the bronchioles of

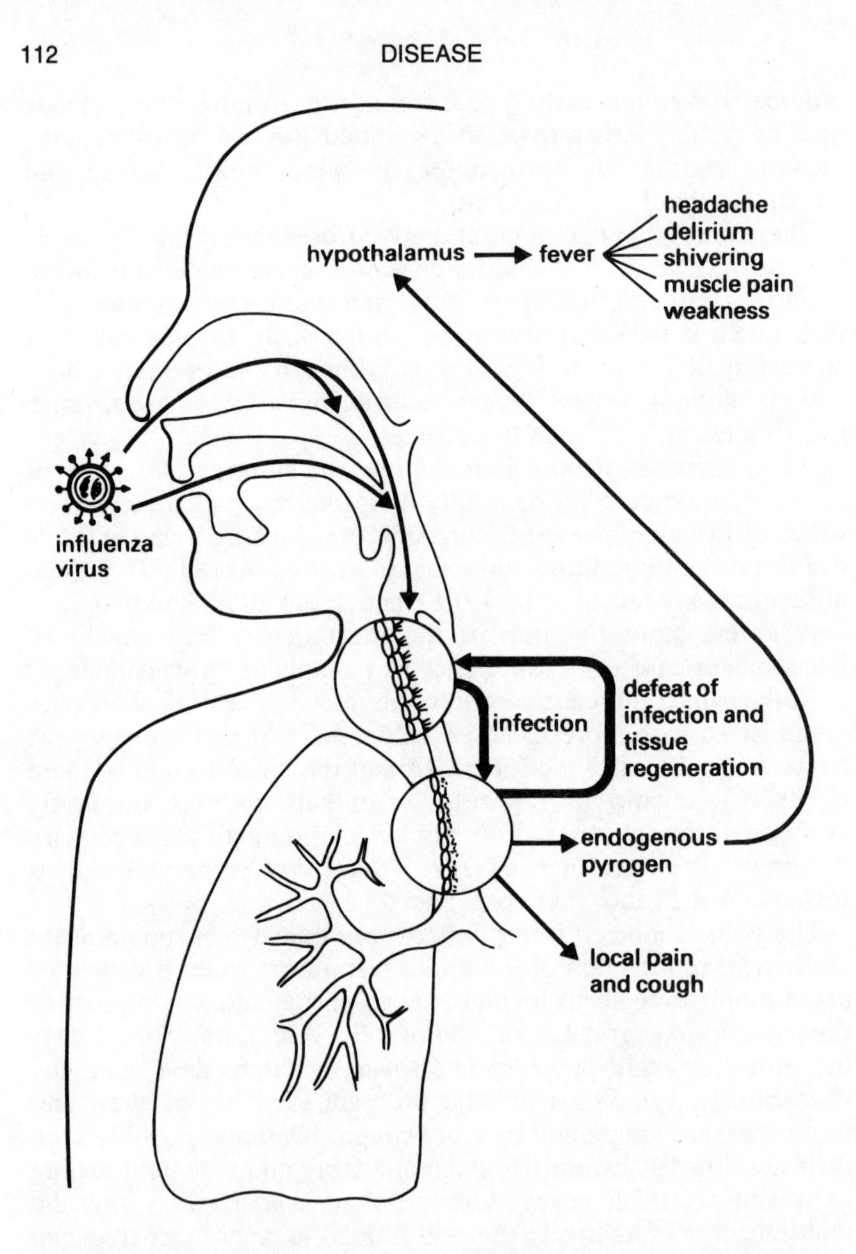

Figure 9.2 The major details of a typical influenza virus infection

the lung and then down into the oesophagus. The oesophagus then carries the mucus to the stomach where many viruses and other micro-organisms carried along by the mucus will be destroyed. Influenza virus infection kills the cilia-bearing cells, causing them to distintegrate and expose the layer of cells directly beneath the ciliated layer (which are not vulnerable to the virus). This damage caused by influenza virus is vividly illustrated by the photographs of plate 9.1.

The throat and chest pains, and also the cough, are a direct result of destruction of the cells that serve as hosts for influenza virus multiplication. But how are these symptoms actually produced? Throughout our bodies there are many specialised nerve fibres known as "pain receptors". If the pain receptors are stimulated to send nerve impulses to the brain then we feel pain, and such stimulation is caused by tissue damage near the receptors, swelling or mechanical deformation of the surrounding tissue, or extremes of heat or cold. Damaged tissue, such as the cells destroyed by influenza virus, releases specific substances that permeate through to the pain receptors and lower the threshold of stimuli needed for a receptor to "fire" (i.e. tell the brain it feels pain). In this way the infection allows previously harmless stimuli to produce pain.

Of course the infected tissue will also become inflamed, resulting in considerable swelling which will put pressure on the pain receptors and again produce pain. So the *local* pain experienced during a bout of influenza – the pain directly due to viral multiplication – can be satisfactorily explained. A similar tale lies behind the pain produced at the site of many different viral infections. As far as the cough experienced during some bouts of influenza is concerned, we also have specialised nerve endings called "cough receptors" dispersed throughout the nose and throat. These can also be stimulated by tissue damage and swelling; hence the cough.

The most miserable features of an influenza attack are the headaches, muscle pains, shivering, weakness and delirium that combine to make the flu a much more wretched experience than other respiratory infections such as the common cold. Precise explanations for each of these symptoms are currently unavailable, but many are at least partly caused by the *fever* that accompanies almost all attacks.

Fever which simply means a raising of the body temperature above its normal healthy level, is a common response to many infectious diseases. The problems it brings arise because the organs of the body, particularly the brain, are very sensitive to changes in temperature. Any deviation from the 37° centigrade mark (at which all the enzymes are working at the right speeds and the correct balance is being maintained between tissue degradation and renewal), can eventually have very serious consequences. A rise to above 40.5° can cause disorientation, delirium and mania; while temperatures above 43.4° can induce coma and death. Cold is also dangerous, with unconsciousness setting in below about 27.7°. The fever associated with influenza usually takes body temperature to somewhere between the 38° and 41° marks, and once again we face the question of how can viral multiplication bring such a change about?

All fevers appear to be caused by a small protein native to the body, known as "endogenous pyrogen", which is produced by various white blood cells in response to infection. Some phagocytes, for example, release it when they engulf a suitable micro-organism; while some of the proteins released from activated T-cells can induce other white blood cells to release endogenous pyrogen. Once it has been released, endogenous pyrogen is carried through the circulation to a part of the brain known as the hypothalamus (see figure 9.2), responsible for regulating body temperature (amongst other things). There are many ways in which the temperature of the body can be controlled. Sweating and the widening of capillaries in the skin, for example, encourages the loss of heat to the environment; while narrowing of the capillaries to reduce heat losses, or shivering to generate waste heat from muscular activity, can make us warmer. In a healthy person the nervous system integrates all the various temperature-regulating mechanisms to keep the temperature around 37°. The effect of endogenous pyrogen is to re-set the "thermostat" in the hypothalamus, turning it up a few degrees to produce a fever. The details of exactly how the thermostat is re-set remain unclear, but the increased production of chemicals known as "prostaglandins" is apparently required. Aspirin and several other anti-fever drugs can interfere with prostaglandin manufacture, which probably explains why aspirin can lower the fever that accompanies influenza.

Having described how infection with the influenza virus manages to bring about a temperature rise, the problem remains of *why* the body should itself produce a potentially damaging fever in response to viral attack. It is an appealing and extremely popular notion that fever somehow "helps" the body to fight off an infection. For example, the sensitive chemical structure of a virus might be disrupted at the higher temperatures, or a speeding-up of body metabolism in general might also speed up the elimination of viruses. Despite the obvious attraction of such ideas there is actually little evidence to back them up. If fever is controlled with drugs it does not seem to take any longer for an infection to be overcome, leading some scientists to suggest that fever is simply an accidental and unfortunate accompaniment to disease. This view is vigorously contested by others who feel that such a drastic and potentially damaging process must have proved of crucial benefit at some point in evolution, otherwise it would not exist. Bearing this in mind, it seems possible that fever is a legacy of evolution which *used to be* of great importance to our distant non-human ancestors, but which is now of little use. Such a change in the status of fever could be explained by improvements in our other defences which have made the fever response redundant; or perhaps by the evolution of hardier micro-organisms more resistant to slight increases in body temperature.

Whatever the truth about its usefulness, the immediate effect of the onset of fever during influenza is not to make us feel better, but rather to make us feel very ill indeed. Headaches associated with fever can be caused by the widening of blood vessels in the brain, increasing the supply of "hot" blood while at the same time putting pressure on pain receptors nearby. Delirium is probably a direct consequence of the damage that the temperature increase does to the overall working of the brain, and similar detrimental effects of increased temperature elsewhere in the body are probably at least part of the explanation for muscle pains and the general feeling of weakness and lethargy. Shivering, of course, is just one of the mechanisms the body uses to produce the additional heat required to maintain a fever.

So we have covered the main features that allow the infection to get established and begin producing the characteristic symptoms of flu. Thankfully, our defences are usually highly effective against influenza virus infections. Anti-viral antibodies appear in the blood only 4 days after the infection first begins, and by the 5th day the virus is usually succumbing to our combined defensive armory and the layer of ciliated cells is beginning to regenerate (see plate 9.1). This recovery is of course accompanied by the subsidence of symptoms, which have usually disappeared completely within 7–10 days. Two weeks after the virus first invaded, the infection has usually been completely repulsed, the ciliated cell layer will be complete once more and the immune response will be on the wane.

Unfortunately this happy picture of a complete and uncomplicated recovery does not always apply. Occasionally serious complications can set in. Pneumonia is the greatest danger, produced when the influenza virus itself penetrates deep into the lungs, or else by bacteria whose entry into the lungs is made much easier by the destruction of the protective mucus layer in the airways above. Whether of viral or bacterial origin, pneumonia is of course an extremely serious illness that can quickly kill, especially in the old or very young. When an influenza infection leads on to death it is usually the secondary pneumonia that is actually the killer.

Other less common complications can arise if the infection manages to spread from the respiratory system to other parts of the body. Infection of the brain, muscles or heart can occasionally produce encephalitis, myositis or myocarditis respectively. But in most cases influenza follows the course of a typical acute viral infection – there is a brief bout of illness followed by a quick and uncomplicated recovery.

We cannot leave the subject of influenza without mentioning the famous tendency of influenza viruses to strike in waves of pandemics and epidemics against which previous attacks of influenza offer little protection. From our discussion of immunity you might expect that one episode of influenza might protect you against subsequent infections for life. Unfortunately this is not the case. First of all, there are several different types of influenza virus around; but most of our problems are caused by the unfortunate ability of influenza viruses to change from one form, against which much of the population might well be immune, into another form, which most of the population will never have encountered. Put simply, the antigens on the surface of influenza viruses are extremely variable, allowing successive waves of viruses to arise carrying antigens that your immune system has never seen before. Some of the details of how the necessary variations come about can be found in box 9A.

Our look at influenza should have provided you with a good insight into the course of many illnesses caused by acute viral infections. Of course influenza is not a perfect example; the almost infinite variability of viral infection means that no examples can ever be perfect, but the story of many other acute infections is similar to the story of influenza. The details of which organs are

BOX 9A – HOW INFLUENZA VIRUSES VARY

There are actually three main types of influenza virus – types A, B and C. The type A viruses cause the most severe epidemics in humans and it is to type A that the discussion in the main text specifically applies. Nevertheless, type B epidemics do occur, while type C causes only a mild cold-like illness without high fever. Type A is the most adept at undergoing the variations that baffle the immune system, and viruses of this type have been responsible for all the major waves of influenza this century.

To understand how the influenza viruses manage to vary you must consider again the structure of an influenza virus overall (see figure 9.3). The viral genome consists of eight separate segments of RNA; while the parts of the virus that come into contact with the immune system, the viral *antigens* in other words, are the two different types of glycoprotein found sticking out from the viral envelope. One of these proteins is called "neuraminidase" (N), while the other is known as "haemagglutinin" (H). The H protein is the one that actually binds to the receptors on the surface of host cells, while N is an enzyme which, amongst other things, might break down mucus.

Both the H and N proteins can vary slightly through the years, due to the slow accumulation of mutations in the genes that code for them. This process of slow gradual change is called "antigenic drift", producing slight but significant changes in the H and N proteins (see figure 9.3). Obviously when these proteins change in structure then the parts of them that serve as antigens may also change, making immunity built up against their earlier forms either much less effective or completely useless. Change in influenza viruses due to antigenic drift is believed to explain the local and relatively mild influenza epidemics that appear every few years. The devastating worldwide pandemics that occur much less frequently are blamed on a much more drastic phenomenon known as "antigenic shift".

Antigenic shift is made possible by the existence within the influenza virus population of several radically different versions of the H and N proteins. At least 13 variants of H are known (labelled H1–H13), accompanied by at least nine variants of N (N1–N9).[2] During antigenic shift the predominant forms of these two proteins found in the viruses infecting humans suddenly change. Prior to 1968, for example, H2N2 was the predominant arrangement and the world's population had become relatively well protected by immunity to these specific proteins. But in 1968

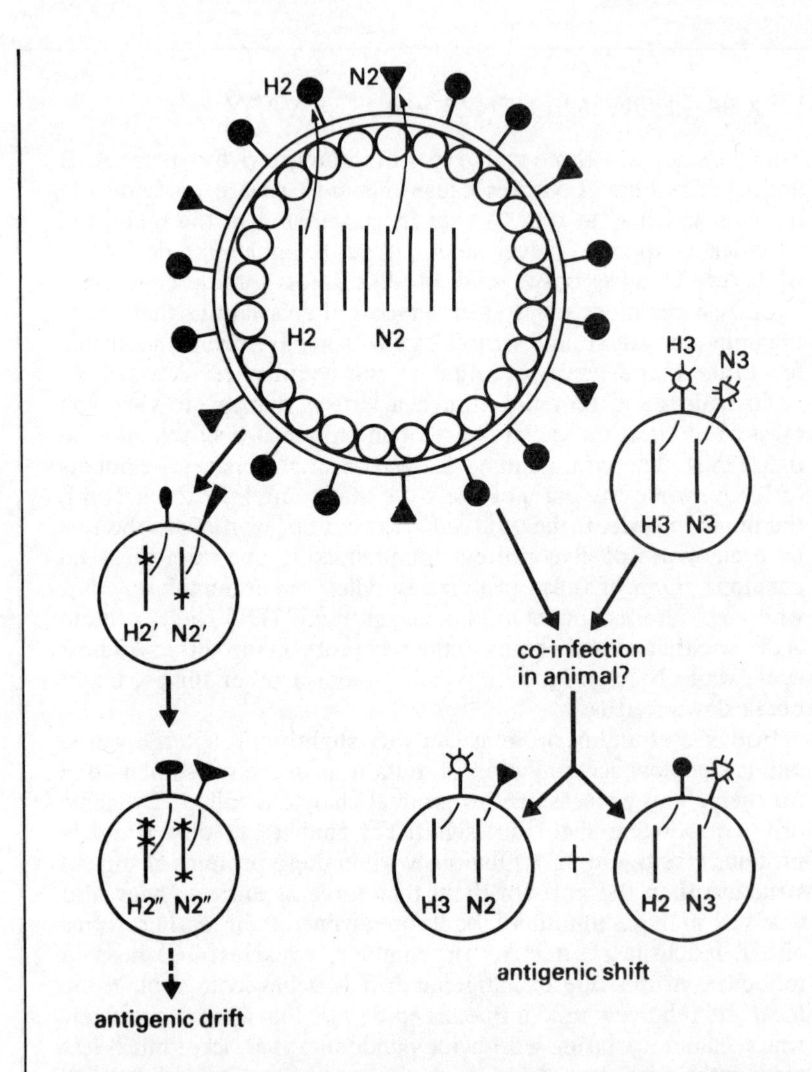

Figure 9.3 The antigenic variation of influenza A viruses. The viral antigens are parts of the "H" and "N" glycoproteins. Mutations in the genes encoding H and N produce slight changes in the structure of these proteins. This process of "antigenic drift" is believed to generate the variant viruses responsible for influenza epidemics. In "antigenic shift" different viruses are believed to infect the same cell and swap RNA segments coding for radically different versions of H and N. The viruses emerging from such swapping are blamed for influenza pandemics. (In the smaller diagrams, only the RNA segments encoding H and N are shown, for simplicity.)

a new H3N2 virus appeared in China and South East Asia. The H3 protein of this virus was considerably different from the H2 found in the earlier form; most people were therefore not immune to it, and so a pandemic of "Hong Kong Flu" circled the globe killing many people on its way. A further pandemic appeared in 1977 when an H1N1 variant surfaced, and with immunity against this form now firmly established we await the next shift that will set the pandemic cycle in motion once again.

Uncertainty still surrounds exactly *how* the new variants that give rise to pandemics are produced, but the favoured explanation points to the segmented nature of the viral genome and the fact that the H and N proteins are each encoded by a separate RNA segment. Suppose two different forms of the virus were to infect the same cell simultaneously. As shown in figure 9.3, the viruses might be able to swap the crucial gene segments between themselves to produce new variant viruses. Such swapping could simply be the result of random packaging of the segments available within the cell into new viruses.

A further problem, however, is presented by the occasional appearance of versions of the H and N proteins that have never been identified in previous human influenza infections. This is explained by assuming that the human viruses can also infect other animals, allowing them to pick up the genes for new H and N proteins from the influenza viruses that infect these other animals. The favourite animal suspected of acting as a "reservoir" of influenza variation in this way is the Chinese duck, which harbours a wide variety of influenza viruses within its intestine. There is an enormous population of ducks in China, where they are encouraged because of their appetite for various pests that attack rice crops (the ducks actually outnumber the Chinese!) By living in close proximity to the large rural population of China, the ducks must present an ideal reservoir from which human influenza viruses can pick up new genes. This proposal also neatly explains why the last three influenza pandemics (1977, 1968 and 1957) have originated in China. I should emphasise, however, that the source (or sources) of influenza virus variation is certainly not firmly established.

infected, exactly what symptoms are produced and how quickly the infection is repulsed will obviously vary; but the overall picture of a burst of virus multiplication and tissue damage, followed by recovery or death, will often be the same. That other well-known acute infection of the respiratory system – the common cold – actually follows the influenza pathway quite closely, with the exception that high fever is rarely produced. This makes the cold a much more trivial illness which many sufferers manage to "battle through" without even taking time off work. Battling through a severe attack of influenza is rarely feasible.

Persistent infection[3]

Medical interest in viral infection has traditionally been focused on those acute infections that in turn produce acute disease. The reasons for this bias are obvious: acute infectious diseases are the most dramatic and easy to identify, and until the era of vaccination arrived they were certainly the most important. But over the past few decades interest has increased in *persistent* infections in which viruses, or at least some part of the viruses, remain in the body for anything from a few months to a lifetime. The course of all viral infections is determined by a complex struggle between the virus and the host's defences. Persistent infections are often the result of stalemate in this struggle. They occupy the middle ground between acute life-threatening infections that can quickly overcome our defences; and infections that readily succumb to these defences, albeit after a short burst of disease.

The details of what is going on during viral persistence are still shrouded in considerable mystery. Various general ways in which persistence might arise are fairly well established and understood (see figure 9.4), but which mechanisms apply to which particular diseases is often unclear.

Viral infections are often able to persist if the immune response to the infection is in some way inadequate. The possible routes towards such inadequacy are many and varied. A generalised deficiency of the immune system accompanies many diseases, is found in fetuses and newborn infants before the immune system is fully mature, and sets in gradually with old age. Immune deficiency can also be induced by exposure to radiation during anti-cancer therapy or by treatment with certain drugs. Some viruses are also themselves able to induce immune deficiency by infecting and interfering with the activity of white blood cells. In all of these situations the persistence of viruses that would normally be quickly fought off is encouraged.

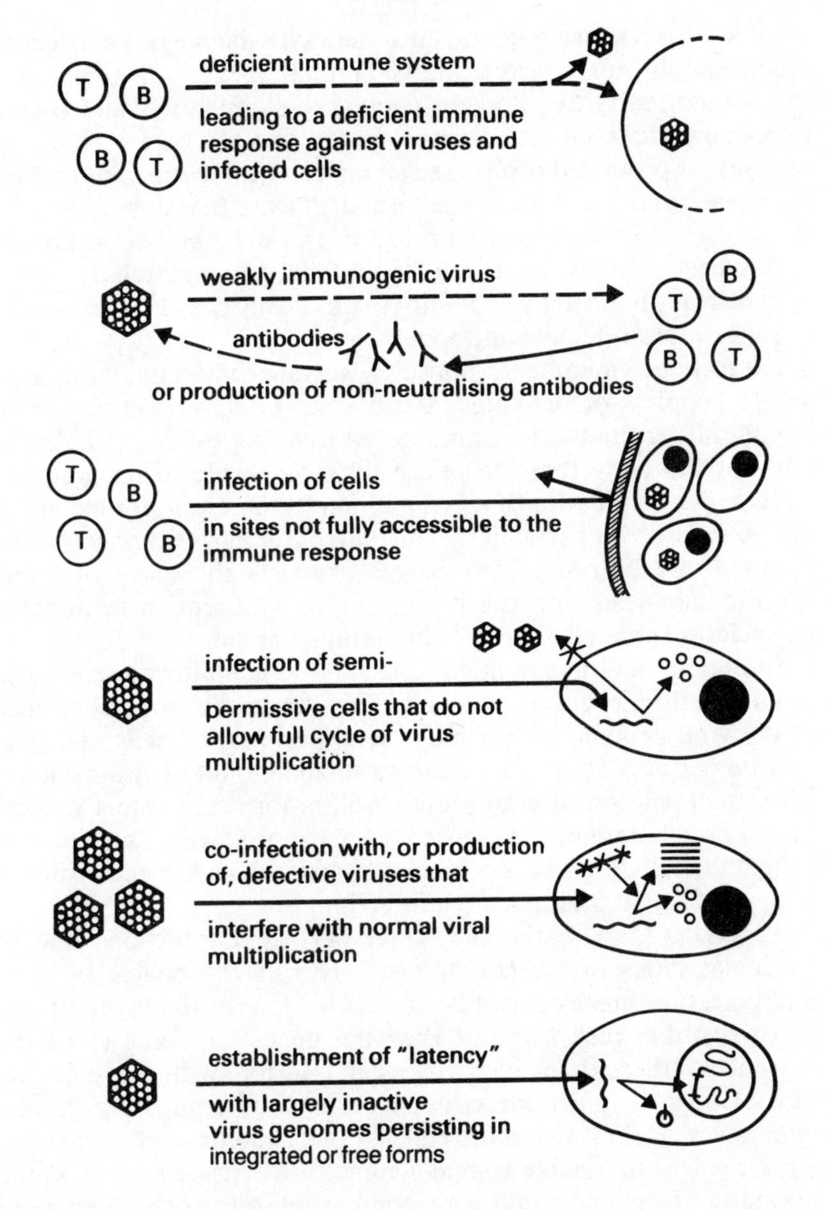

Figure 9.4 Some mechanisms of viral persistence (see text for details)

Even in healthy adults, however, the immune response against a virus can sometimes prove inadequate to eliminate the virus altogether, while being sufficiently effective to prevent the infection from causing death. In some cases the viruses concerned might simply not be particularly "immunogenic", meaning that they do not carry the antigens needed to stimulate T- and B-cells into a powerful immune response. But even when a strong immune response *is* produced problems can arise. The resulting antibodies, for example, might bind to the viruses in ways that do not prevent the viruses from infecting their host cells. Such "non-neutralising" antibodies certainly exist, and their failure to neutralise a virus may be compounded by an ability to physically prevent the binding of other more effective antibodies.

Even if the immune response is working perfectly, producing ample supplies of activated T-cells and effective antibodies, it might still be unable to eliminate viruses that become hidden in parts of the body that are relatively inaccessible to the immune system. The capillaries that bring blood to the brain, for example, are bounded by a particularly restrictive cell barrier known as the "blood-brain barrier". This barrier restricts the entry of many specific chemicals into the brain and also presents a formidable obstacle to some elements of the immune system.

Sometimes a virus can infect cells that do not allow the full cycle of viral multiplication to proceed. This can lead to the persistence of the viral genome within such "semi-permissive" cells, where it may be partially active, producing a small amount of viral protein. This small amount of viral protein will in turn stimulate a weaker than normal immune response against it, allowing the virus to escape the full defensive onslaught that would be directed against a fully established multiplicative infection.

A similar tale lies behind some persistent infections due to particular types of *defective viruses*. During the course of virus multiplication mistakes can be made, leading to the manufacture of viruses that lack some of the viral genes. In some cases the presence of these defective viruses can interfere with the manufacture of normal infectious viruses, perhaps by competing for the vital raw materials within the cell. At the same time, the defective viruses might be unable to make some of the proteins that would normally trigger the immune response, and so the overall effect of the appearance of defective viruses might be a weakened infection accompanied by a weakened immune response – all of which could lead to persistence.

Finally, various viruses can persist thanks to a phenomenon known as "latency", which bears a close resemblance to the "lysogeny" discussed in chapter 5. In lysogeny, remember, bacteriophages were able to persist within a bacterium by integrating their genetic material into the bacterial genome. Once integrated, the viral genes were quietly replicated along with the bacterial genes while remaining in a largely inactive form. The genetic material of some viruses that infect humans and other animals can also persist in a largely inactive "latent" form. In some cases such latent genomes do integrate into the host cell DNA, while in other cases they may remain as independent quiescent viral genetic material within the cell. Just like bacteriophage genomes in lysogeny, latent animal virus genomes may become activated at a later date to initiate a new bout of viral multiplication.

In all forms of persistent infection the presence of the persistent virus must obviously not be too damaging or the infected host would quickly die. Sometimes the viruses concerned are simply not particularly harmful, even during full-blown multiplicative infections. In other cases the effects of a potentially damaging virus are still significantly limited by a weakened or simply ineffective immune response. Latent viruses, of course, are in a largely inactive state which might produce neither damage nor much of an immune response. The eventual illnesses associated with viral persistence may be caused by the slow accumulation of chronic tissue damage brought about either by the infections themselves or the victims' immune responses.

So much for generalisations about persistent infections; what about the specific diseases they cause? The progress of an illness brought about by viral persistence can, you will not be surprised to hear, follow many different paths. Some of these paths are summarised in figure 9.5, which is again an attempt to bring some order and simplicity to what is really an infinitely variable process. The figure illustrates that some persistent infections cause periodic outbreaks of illness over a long period of time; some produce a first burst of illness and then lie quiescent for years before re-surfacing in a different form; some initiate a gradual build-up of illness to a peak; while others produce chronic illnesses from the outset. Let's consider a few examples.

Genital herpes[4]

No single viral infection has received more publicity in recent years than genital herpes – a recurrent sexually transmitted infection of the genitalia. The disease is reported to have flourished in the post-pill era of sexual promiscuity, to such an extent that *fear* of the disease may be acting as an effective brake on that promiscuity. To some extent, however, the fuss over genital herpes is a media creation. It still accounts for only a small percentage of attendances at genito-urinary clinics swamped by many more cases of other sexually transmitted infections.

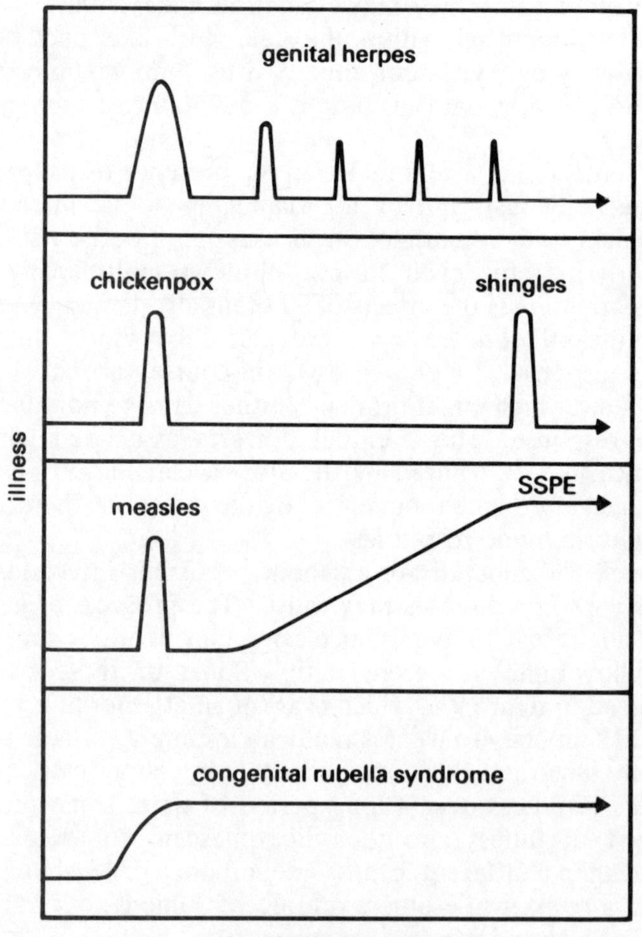

Figure 9.5 The development of various diseases involving persistent viral infection, illustrating some of the alternative consequences of persistence

The herpesviruses are, as already mentioned, a large and varied group of viruses which cause many different diseases; but when most people talk of "herpes" they mean the genital infection caused primarily by "herpes simplex virus type II". A closely related virus, herpes simplex type I, causes cold sores around the mouth and is increasingly being found in genital infections as well.

In genital herpes the virus infects the cells surrounding and lining the genitalia and quickly establishes an acute infection. The majority of cases are probably inapparent; but if the infection really gets a hold it produces the localised tissue destruction, inflammation, pain and possibly fever associated with a first attack of "herpes". This initial acute infection soon subsides as the body mounts a seemingly very effective immune reponse, but of course the story does not end there. Having produced one bout of illness, the infection can then recur at intervals ranging from a few weeks to over a year. Each recurrence yields similar symptoms to the first attack, although usually much less severe. The source of the recurrences is of course a persistent infection which the immune system cannot eliminate, but instead of staying in the cells that were initially infected the virus persists in nerve cells far from the infection's initial site. As the immune system overcomes the acute infection around the genitalia, the viruses can enter the nervous system and "retreat" along the nerve fibres to large bundles of nerve cells known as "ganglia". Within the cells of these ganglia, probably close to the spine, a latent infection is set up. The viral genes become largely inactive, producing little or no viral protein and certainly not manufacturing complete infectious viruses. The state of the herpes genome during latency remains unclear – it might integrate into cellular DNA or it may persist independently – but in either case it remains quietly awaiting the opportunity to become active once more. When re-activation does occur the virus appears to multiply briefly in the nerve cells, and then travels back along the nerve fibres to the initial site of infection. There it establishes another bout of acute infection in the same type of cells as were infected in the first place.

Sufferers from genital herpes would obviously like to know what *causes* reactivation of the latent virus and what can be done to prevent it. Many factors that encourage re-activation have been identified but the ways in which they do so are not yet known in any detail. The identified risk factors likely to lead to re-activation include injury to the affected tissue (even the relatively mild irritation associated with intercourse), exposure to strong sunlight or other sources of ultra-violet radiation, contact with chemical irritants, menstruation, illness in general, fever, immune suppression and even anxiety, depression and stress. Many of these stimuli may work in similar ways. Depression, for example, has been shown to diminish the effectiveness of the immune system. Whatever the explanation they should obviously all be avoided as far as possible. Translated into simple practical terms this means restrict sexual activity, avoid nude sunbathing (the two may be related!), avoid irritating soaps and sprays, eat well and try to keep fit, healthy and happy. Such precautions are certainly not a guarantee against recurrence, but they should reduce both its frequency and severity.

Ideally, of course, effective drugs are needed to conquer or at least control the disease. In recent years the increasing success of research towards anti-viral drugs has been particularly promising in the area of anti-herpesvirus therapy. We will be looking at the relevant drugs and research in chapter 12. In the meantime, another member of the herpesvirus group can be used to illustrate a different pattern of disease due to persistent infection – an initial bout of acute illness followed by a somewhat different illness some years later.

Chickenpox and shingles[4]

Chickenpox, one of the most common viral infections of childhood, is caused by a herpesvirus known as "varicella-zoster virus" (varicella is the medical term for chickenpox, while zoster is the term for shingles). This virus enters via the respiratory system, multiplies, and then gains access to the blood. The bloodstream carries it throughout the body to set up acute infections at many sites. The dramatic rash is due to infection of the skin and an associated inflammation.

The initial infection is usually defeated within a couple of weeks, but as with herpes simplex the disappearance of symptoms does not mean that the virus has been completely eliminated from the body. It too can apparently enter the nervous system and remain latent for years. Later in life the virus can emerge from its dormant state, travel back down the nerves and cause the skin rash, fever and pain that we call "shingles". Unlike herpes simplex, varicella-zoster virus rarely causes multiple recurrences of disease. Most people who reach the age of 80 will have suffered one bout of shingles, but further attacks are unusual. Again, exactly what causes the latent infection to be re-activated is not known. Various observations have encouraged the belief that the latent infection is held in check by a healthy immune system, but can re-emerge as a multiplicative infection when immunity is impaired. First of all, shingles appears in old age, when the effectiveness of the immune system is known to be on the wane; and secondly, the disease often occurs alongside cancer (which can be associated with immune deficiency) or in patients treated with immunosuppressive drugs.

Measles and SSPE[5]

Another common childhood infection – measles – provides us with an example of a quite different pattern of persistence and disease. This time the consequences of the persistent infection build up slowly over a long period of time, rather than causing two or more bursts of acute disease separating periods of apparent good health. Measles virus normally causes an acute infection that is similar in many ways to chickenpox. The virus again enters via the respiratory system, multiplies briefly in the cells lining the respiratory tract, and then spreads throughout the body via the lymphatic system and the blood. The virus can infect a wide variety of organs such as the skin, liver, tonsils, spleen, lungs and blood. Once again it is infection of the skin that produces the most obvious sign of disease – a rash that begins at the forehead and then spreads downwards to the feet. The infection is normally soon defeated but occasionally the virus can spread to the brain to cause encephalitis and perhaps death. This ability to spread to the brain allowed measles virus to kill significant numbers of children prior to modern vaccination campaigns.

So far we have looked only at the effects of *acute* measles virus infection. In a small number of cases the initial acute infection is apparently fought off (or might never be noticed), only for a slowly progressing and potentially fatal encephalitis to emerge some years later. This unusual consequence of measles (found in only 1 out of every 200,000 cases) is known as "subacute sclerosing pan-encephalitis (SSPE). It usually occurs in children who catch measles before they are 2 years old, with the eventual disease becoming evident between the ages of 4 and 18. Initially the effects are very mild – the child's parents might notice a lack of concentration and teachers may complain of deteriorating schoolwork; but the symptoms then slowly worsen through phases of more obvious unusual behaviour into erratic jerking of the limbs, blindness, severe mental retardation and then death.

SSPE is caused by a persistent measles virus infection of the brain, leading to the death of nerve cells and "demyelination" (the break-up of the fatty sheath around nerve fibres which is essential for them to work properly). It seems likely that several of the mechanisms of persistence discussed earlier might be involved. First of all, the virus isolated from SSPE victims appears to be defective. Secondly, the infected cells are possibly "semi-permissive", preventing the virus from multiplying properly even if it is not defective. Thirdly, a defective immune response is probably raised against the virus, perhaps because a defective virus within semi-permissive cells is not producing the viral proteins needed to trigger full immunity. Much of the actual damage caused by the persistent infection could well be due to an autoimmune response set up against the infected brain. Autoimmune antibodies *have* been found in SSPE victims, and we saw earlier that some of the measles virus proteins are apparently very similar to proteins found in healthy human cells. Fortunately, the incidence of SSPE has considerably declined following the widespread use of anti-measles vaccines, but research into the disease as a model of viral persistence in general will undoubtedly continue.

Congenital rubella syndrome[6]

We have looked, then, at persistent infections responsible for recurrent outbreaks of disease, a single acute resurgence after many years, or a slow build-up of illness following a period in which the initial infection had apparently been fought off. In our final example, "congenital rubella syndrome", the virus multiplies continually throughout the diseased tissue to produce a chronic infection with disastrous results.

Rubella, best known as "German measles", is normally a harmless measles-like infection that produces a rash for a few days and then is gone. It is probably transmitted by the respiratory route, followed by entry into the blood and a trivial infection spread throughout the body. Our immune defences can overcome the infection with ease, but problems of persistence arise if it spreads from a pregnant woman into the vulnerable tissues of her unborn child. The devastating consequences of rubella infection in a fetus are paradoxically partly due to the relative harmlessness of the virus. A fetus is very poorly protected against infection because its immune system does not start to work properly until just before birth at the earliest. This vulnerability is partly alleviated by antibodies passed on from the mother's bloodstream, but these provide only limited protection at best. The result of fetal vulnerability is that acute viral infections that cause serious cell damage usually kill a fetus and the pregnancy aborts. Only relatively mild infections, such as rubella, allow a fetus to survive through to birth; and although the effects of the virus may be mild from the point of view of individual cells, they can seriously interfere with the intricate programme of tissue development that changes a fetus into a properly formed child.

A persistent chronic rubella infection can spread throughout all of the developing fetal organs, interfering with cell division and producing deformities such as blindness, deafness, mental retardation and defects of the heart, bones and blood. About 15 per cent of rubella-infected babies die before they are 1 year old. In those that survive the infection is eliminated as soon as the immune system becomes fully operational, but by then it is far too late. Fortunately, rubella is another illness that can be prevented by vaccination. The development and use of anti-rubella vaccines would never have been justified by the mild and often unnoticed symptoms the virus produces in adults, but they are of course amply justified by the horrific effects of the virus on unborn children.

Serum hepatitis[7]

Our rapid journey through the complex landscape of persistent infection should have given you a good overall impression of that perplexing area of virology. It leaves our review of the various patterns of disease caused by the viruses almost complete (apart from cancer). We have looked at inapparent infections that cause no noticeable illness, considered the progress of one particular acute infection in some detail and examined some of the mechanisms and consequences of viral persistence. But in many cases an individual type of virus can follow any of these paths, with the choice of route being made by poorly understood influences such as our general state of health, the amount of viruses we are exposed to and of course the levels of any pre-existing immunity. To round this chapter off it will be worthwhile to look very quickly at the alternative paths available to one specific virus, and the frequency with which the various paths are followed.

Serum hepatitis is the most common form of hepatitis (inflammation of the liver) and is caused by a complex double-shelled "hepatitis B" virus which carries a DNA genome. The virus is commonly passed from person to person in contaminated blood (putting hospital workers, dentists, laboratory staff and drug addicts at particular risk), but it can also be transmitted in infected saliva, urine and semen. Once the virus gains access to the bloodstream it proceeds to infect cells throughout the body, but particularly in the liver. Inapparent infection is the most common result (see figure 9.6), but in about 35 per cent of cases an acute multiplicative infection in the liver can cause hepatitis, jaundice (as a result of the liver damage) and fever. About 3 per cent of the acute apparent infections will turn out to be fatal; but the most troublesome aspect of hepatitis B infection overall is the persistent infection seen in about 5 per cent of the virus's victims. The incidence of persistence actually varies widely throughout the world, being comparatively rare in the West but found in up to 20 per cent of cases in underdeveloped countries. Many of these persistent infections are passed on from mother to child at birth, via contact with infected maternal blood.

Exactly what allows persistence to set in in the first place is not clear, but an inefficient immune response certainly seems to be a factor. For example, although high levels of anti-viral antibodies are found in the blood of people with persistent infections, one crucial class of antibody (directed against the surface of the virus) is usually lacking. Persistence can set in after an attack of acute hepatitis or following an otherwise inapparent infection, and all persistently infected individuals become "carriers" of the disease, able to pass it on to other people.

There are likely to be at least 150 million hepatitis B carriers dispersed throughout the world, making them a major factor in the spread of the more obvious acute infection. So-called "healthy" carriers, in whom the persistence causes no liver disease, only have to worry about passing their infection on to someone else; but in other less fortunate carriers the persistent infection can cause chronic liver disease, liver failure, or perhaps cancer.

Mention of the link between persistent hepatitis B infection and cancer leads us neatly into the whole fascinating topic of virally induced cancers. That is a subject not for this chapter, but for the next.

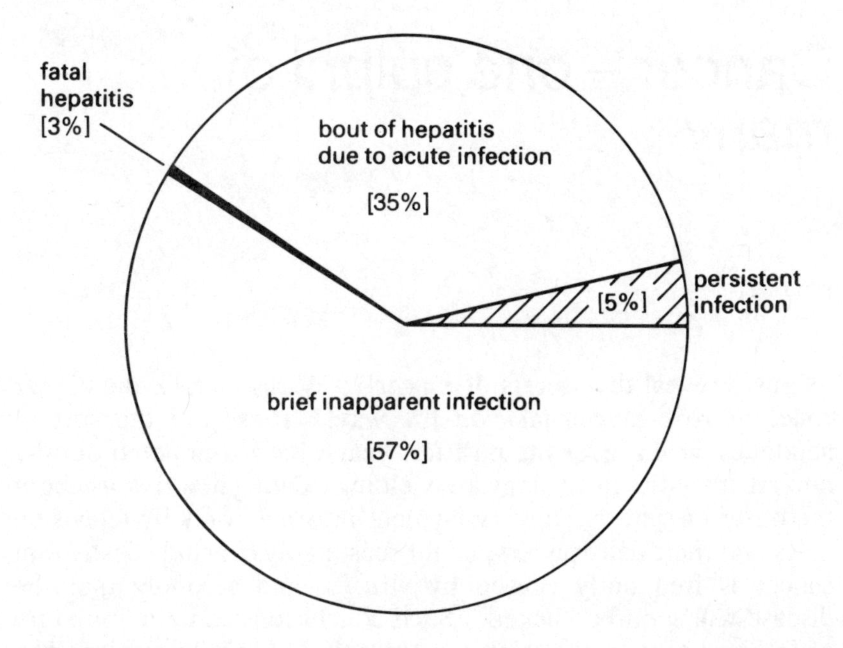

Figure 9.6 Outcomes of infection with hepatitis B virus (figures approximate)

CHAPTER 10

Cancer – one culprit of many

"Viruses reveal the secrets of cancer!"; "Viruses crack the cancer code!"; "Anti-cancer jabs on the way!" These are the sort of headlines which, over the past few years, have trumpeted out the news that studies in virology are yielding exciting new details about the onset of cancer. Unwary laymen, informed only by television news and their daily papers, might reasonably conclude firstly that cancer is frequently caused by viruses, and secondly that the disease will soon be "licked". Such conclusions contain a mixture of fact and fantasy which this chapter will hopefully resolve. The enthusiasm with which scientists are now embracing the cause of cancer viruses might suggest to outsiders that the field is a very new one. In fact, the first links between viruses and cancer were actually established over 75 years ago, only 10 years after Beijerinck discovered his mysterious "soluble living germ".

Chickens and mice

During the early years of this century a doctor called Vilhelm Ellerman and a vet called Olaf Bang were working together on chicken leukaemia, a disease which in those days was not even recognised to be a form of cancer. The two colleagues soon discovered that they could transmit the disease to healthy chickens using filtered blood taken from a diseased chicken. In retrospect, we can identify this discovery as the earliest evidence linking the viruses with cancer, but when it was published in 1908 it received little attention.[1]

A bit more attention was paid to a similar discovery announced in 1911 by Peyton Rous of America, this time concerning a recognised solid cancer of chickens. Rous cut out the cancerous tissue from diseased chicken breasts, added it to mineral solution and then ground it up with sand. The resulting cell debris was then removed by centrifugation and the remaining fluid portion passed through a filter. The filtered fluid was completely free of cancer cells, and yet when it was injected into healthy chickens they often went on to develop cancer.[2] At the time, Rous did not realise that the infectious agent responsible for transmitting the cancer was a virus, but that fact slowly became apparent over the coming years. A British doctor called William Gye was so impressed by this discovery that he wrote a book optimistically entitled '*The Cause of Cancer*' declaring his belief that cancer was essentially a viral disease.[3] Unfortunately, Gye's belief that the mystery of cancer had been solved was hopelessly premature.

By the mid-1930s, however, Rous and other scientists had shown that quite a few animal cancers were transmissible using filtrates of virus-infected tissue; but scepticism and disinterest remained the most common response to the emerging "virus theory" of cancer. Much of the early reluctance to believe in the viruses as a cause of cancer probably stemmed from a simplistic desire to find just *one* cause of *all* cancers. If viruses were to be accepted as that cause then cancer would need to behave like a typical virus infection. This was obviously not the case, especially considering the lack of any obvious pattern of person-to-person transmission.

The disinterest in the viruses as carcinogens began to be overcome in 1936, when an American biologist called John Bittner demonstrated for the first time that *naturally transmitted* viral infections could also cause cancer. He found that breast cancer could be passed on to newborn mice by an infectious agent carried in their mothers' milk.[4] The agent was eventually identified as a retrovirus and is now known as mouse mammary tumour virus (MMTV).

The discovery of MMTV obviously aroused great interest in the possibility that similar viruses might be involved in human breast cancer, and retroviruses have indeed since been occasionally implicated in that disease. In one study a retrovirus was found in the milk of 60 per cent of women whose families had a history of breast cancer, but in only 4 per cent of women with no family history of the disease. Evidence was also obtained suggesting that a retroviral genome is often present within breast cancer cells themselves.[5] Such results, however, certainly do not prove that a retrovirus actually *causes* human breast cancer. In most people's opinion the question of whether or not viruses really are involved in causing human breast cancer is still open.[6] Certainly there is no evidence suggesting that the cancer can be transmitted by breast-feeding in a similar way to the disease in mice.

Some of the difficulties met when trying to pin down viruses as a cause of human cancers will be considered shortly, but Bittner's early work demonstrated one aspect of the problem: that to seek out a viral cause of any cancer, without taking other contributory factors into consideration, might be a drastically oversimplified approach. He made the crucial observation that the involvement of MMTV in cancer is not a simple story of infection followed by the inevitable development of cancer. Instead, the end result of an infection is dependent on at least two other factors – the genetic make-up of the infected mouse and the prevailing levels of various hormones. The virus will only cause cancer if the overall constitution of an infected mouse is "right".

With Bittner's discoveries belief in the viruses as carcinogens became respectable and research into the phenomenon began to accelerate up to its present hectic pace. A wide range of both retroviruses and DNA viruses are now accepted as undoubted causes of cancer in many types of animal including birds, cats, rodents, monkeys and cattle. These viruses are not just of academic interest, but are capable of causing considerable economic loss, both due to the direct effects of livestock infections and the expensive measures such as vaccination required to control them.[7]

By far the most intriguing and pressing issue raised by the discovery of cancer viruses is whether or not they cause cancer in humans. It might seem naive to expect that we might somehow be protected against a consequence of viral infection to which many other species are clearly vulnerable, but that rather complacent attitude was favoured for quite some time. A few years ago the talk was of "possible" or "suspected" links between viruses and human cancer; but the level of confidence has recently increased sufficiently for Professor Anthony Waterson of London's Royal Postgraduate Medical School to declare "At least some human cancers are caused by viruses", although he quickly added "but not all – and perhaps not even more than a small minority".[8] Let's take a look at some of the evidence justifying this point of view.

Human cancer viruses

Proving that viruses can cause human cancer was always going to be much more difficult than implicating them in animal carcinogenesis, given that the involvement of viruses in cancer is considerably less obvious than their involvement in most typical viral diseases. The ultimate test of any supposed cancer virus is to demonstrate that it really can cause cancer when healthy individuals are infected on purpose. This is possible when laboratory animals are the subjects (although many people fiercely object to such experiments), but is clearly unethical if performed on humans.

There are actually four criteria (known as "Koch's postulates" after Robert Koch, the pioneer of the germ theory of disease) that should ideally be satisfied before any infectious agent can be held responsible for any particular disease. First of all, the suspect micro-organism should always be found in the diseased tissue. This requirement neglects the possibility that the disease might be caused by several different agents, and so it is often weakened to a need to find the micro-organism "regularly" in diseased tissue. Secondly, it should be possible to purify the suspect micro-organism from diseased tissue and grow up a pure culture of it in some suitable environment. Of course, in the case of a virus the suitable environment would need to be an appropriate type of cultured host cell. Koch's third postulate requires that the pure preparation of the micro-organism grown in culture should actually be able to cause the disease when administered to healthy hosts (the ultimate test considered above). Finally, it should be possible to re-isolate the micro-organism from the resulting diseased tissue.

Koch's postulates then, provide a set of *ideal* criteria, but of course both life and science rarely follow the ideal course. Clearly the crucial criterion set by Koch is the one that cannot be met when studying serious human disease – the artificial transmission of the disease to healthy people. Faced with this difficulty, virologists have set themselves a different set of more realistic criteria that should ideally be met by any human cancer virus. Firstly, the incidence of the particular type of cancer in question should obviously correlate with the incidence of infection by the suspect virus. One important piece of evidence pointing to such a correlation would be the presence of antibodies directed against the virus more often, or in greater amounts in patients suffering from the cancer, than in healthy people. Secondly, infection with the virus should be seen to *precede* the onset of cancer, rather than following on from and perhaps being a result of carcinogenesis. Thirdly, the virus or some part of the virus (its genetic material for example) should be consistently found inside the cancer cells. Next, virus purified from the cancer cells should either cause cancer in laboratory animals, or else convert cultured human cells into cancer-like cells. A final and very convincing link implicating a virus in human cancer would be the demonstration that vaccinating people against the virus actually reduced their chances of suffering from the cancer. Any such demonstration would of course bring some reality to the great newspaper headline hope of "anti-cancer jabs".

One very common cancer that could well be the first to succumb to anti-viral vaccination is liver cancer, or more specifically "primary hepatocellular carcinoma". This is the specific cancer found in 80 to 90 per cent of all cases of "liver cancer", and when I speak of liver cancer in the discussion that follows I am referring specifically to primary hepatocellular carcinoma – by far the most common form of the disease. It is one of the top ten cancers world-wide and in much of the Third World it accounts for up to 30 per cent of all cases of cancer. As recently as the early 1970s most medical textbooks made no mention of any possible viral involvement in liver cancer, a situation that has been rapidly transformed into the current view that hepatitis B virus infection is responsible for (or at least crucially involved in) most cases of the disease. This is despite the fact that hepatitis B virus has never been cultivated in cell cultures, making even the direct demonstration that it causes cancer in animals or cultured human cells impossible. So the incrimination of hepatitis B virus in liver cancer provides an excellent example of the indirect methods forced upon virologists in their quest for links between viruses and human cancer.

The earliest evidence against the hepatitis B virus came when several research groups found a strong correlation between the incidence of hepatitis B infection and liver cancer.[9] In the first place, deaths from liver cancer are extremely common in those parts of the world (such as Africa and Asia) in which hepatitis B infection is also common. Secondly, people who die of liver cancer are much more likely than the rest of the population to have suffered from hepatitis B infection early in life, or to be carriers of the disease.

For example, between 1975 and 1978 R. Palmer Beasley of Washington University identified over 3000 male Taiwanese civil servants who were carriers of hepatitis B, and over 19,000 who were non-carriers. In co-operation with Taiwanese scientists he enrolled both groups into a research programme to examine their future incidence of liver cancer. The results were impressive. By the end of 1980, 41 men in the study had contracted and died of liver cancer and all but one of these men came from the hepatitis B carrier group.[10] Even such strong results do not prove that the virus actually *causes* the cancer, but having established that a correlation does exist virologists then went on to examine the relationship between the two diseases more closely.

Many studies have found a considerable excess of both hepatitis B virus antigens and anti-viral antibodies in liver cancer patients, compared to the amounts of these materials in healthy people.[11] Examining the actual cancer cells has proved even more interesting. In 1977 Larry Lutwick and William Robinson of Stanford were able to isolate hepatitis B virus DNA from liver cancer cells,[12] and then a few years later it was shown that the viral DNA was actually integrated into the cancer cells' own chromosomes.[13] The hepatitis B virus genome had apparently become a permanent part of the genetic information carried by most of the cancer cells examined. Carriers of hepatitis B virus were always found to have viral DNA somewhere in their liver cells, and the longer a person had been a carrier the more likely it was that integration of the DNA would have occurred. In some cases viral DNA was also found to be integrated into non-cancerous cells taken from a cancer patient's liver. The usual interpretation of all these results is that integration of the viral genes takes place *before* a liver cell becomes cancerous, and that it is the activity or presence of the integrated genes that actually causes the cancer.

The research summarised so far satisfies many of the criteria required to incriminate hepatitis B virus as a cause of human liver cancer. The virus has been linked to the cancer on epidemiological and geographical grounds; viral proteins and antibodies against these proteins are more prevalent in cancer patients than the population at large; viral genetic material is present within the cancer cells and there is good evidence to suggest that infection with the virus does precede the onset of cancer. The major missing link is the demonstration that the virus can be used to cause cancer in other species or in cultured human cells. This is, of course, an extremely important omission which will hopefully be put right by future research.

Scientific evidence is rarely perfect, but there usually comes a point at which evidence accumulated from diverse sources is generally accepted as making an overwhelming case. The evidence suggesting that hepatitis B virus infection can cause cancer now seems to have reached this stage, with most researchers now being more interested in *how* the virus can give us cancer than in whether or not it really does. With the case against hepatitis B virus considered as proven (being blamed specifically for up to 80 per cent of all cases of primary hepatocellular carcinoma), the virus now stands second only to cigarette smoking in the league table of *known* human carcinogens.[14]

Of course the discovery that hepatitis B virus "causes" primary hepatocellular carcinoma certainly does not mean either that it is the *sole* cause, or that it can produce cancer regardless of other aspects of the overall constitution of its victims. Remember the MMTV virus, which certainly can cause breast cancer in mice, but only if the genetic make-up and hormonal state of an infected mouse allows. Similarly, the consensus about most liver cancers is that they are the cumulative result of several interacting factors, only one of which is hepatitis B infection.[11] Some other factors suspected of involvement are the victim's overall genetic make-up and state of health, the condition of the immune system, the levels of various hormones and exposure to chemical carcinogens. In some cases these other factors might unite to produce cancer without the participation of hepatitis B virus, accounting for those cases of liver cancer that show no evidence of viral involvement.

As research into all types of cancer progresses it is becoming increasingly obvious that it is not a simple disease with one or even just a few distinct causes. Instead, all the evidence suggests that it is produced by a multi-step process, with many different factors available as potential mediators of each step along the road to cancer. In the case of liver cancer, hepatitis B virus certainly seems to be able to take liver cells through one or more steps along that road.

So liver cancer might be largely preventable, if only anti-hepatitis B vaccines can prove to be as spectacularly successful as the vaccines that rid the world of smallpox (see chapter 11). Several different hepatitis B vaccines are now available and the initial steps towards a possible global effort to eradicate the virus are currently under way. If such endeavours significantly reduce the incidence of liver cancer we may well begin to wish that William Gye had got it right, and that all cancers were viral in origin and therefore vulnerable to vaccination. That certainly does not seem to be the case, but there are other human cancers that are probably caused by viruses.

The evidence against some of the other suspected human cancer viruses is even more convincing than that gathered against hepatitis B virus. A herpes virus known as Epstein–Barr virus, for example, as well as meeting many of the other required criteria, will actually produce cancer when injected into monkeys and will transform healthy cultured human cells into cancer-like ones. Epstein–Barr virus (in association with other factors) is believed to produce two rare human cancers known as "Burkitt's lymphoma" and "nasopharyngeal carcinoma".[15]

Specific forms of human leukaemia involving the cancerous growth of T-cells in the blood are now blamed on a retrovirus known as "human T-cell leukaemia virus" (HTLV).[16] Interestingly, the leukaemias concerned have been shown in at least one study to be clustered around specific cities according to the victims' places of birth (rather than where they happen to be living when the cancer arises). This clustering provides evidence of an infectious cause of the cancer and a rather long period of dormancy before the cancer appears. The evidence against HTLV is impressive. The virus is almost always found in the specific type of leukaemic cells involved, while healthy people only occasionally show signs of infection. Anti-HTLV antibodies are much more common in the leukaemia patients than in the population at large. The viral genome is integrated into and active from the chromosomes of cancerous cells, and the purified HTLV virus can change cultured human T-cells into cancerous cells that behave remarkably like the naturally occurring cancer cells isolated from patients.

It is difficult to make any universally acceptable assessment of the current status of viruses as human carcinogens, since differences of opinion always arise in judgements based on large and diverse bodies of evidence. Nevertheless, most virologists now seem to accept that viruses certainly cause some human cancers, with hepatitis B virus, Epstein–Barr virus and HTLV as the most conclusively incriminated examples. Other possible human cancer viruses, against which the evidence is in some cases less firm, are herpes simplex virus type II (strongly suspected of involvement in cancer of the cervix), a herpesvirus called "cytomegalovirus" (which may cause a rare cancer known as Kaposi's sarcoma) and small DNA viruses known as "papillomaviruses" (which certainly cause warts and may cause cancer of the skin, cervix and genitals). Some scientists, however, still prefer to speak of the "possible" involvement of viruses in human cancer, rather than adopting the more confident assessment of Professor Waterson quoted earlier.

If the most important message of this chapter so far is that viruses can sometimes cause human cancer, then the next most important message is that the onset of the cancers concerned almost always involves other factors in addition to viral infection. Only a very small proportion of people infected with hepatitis B virus will ever get liver cancer; only a small proportion of genital herpes sufferers develop cancer of the cervix; and very few people infected with Epstein–Barr virus (which causes glandular fever) ever suffer from Burkitt's lymphoma. The Epstein–Barr virus, for example, only seems to be associated with cancer in certain parts of Africa, and particularly in people also suffering from malaria. People in western countries who have had glandular fever have little or nothing to worry about. A large percentage of the population of the United States has been shown to be infected with Epstein-Barr virus in a latent form, and yet it produces no plague of Burkitt's lymphoma.

The absolutely crucial point emerging from research into viruses and cancer is that the viruses involved should be regarded as *part* of the cause rather than *the* cause of any cancers they are associated with. Professor Murray Gardner has provided a most succinct summary of this situation as follows: "Under natural circumstances tumour viruses can be considered as risk factors which in themselves are neither necessary nor sufficient to produce cancer; they may do so, however, if provided with suitable genetic and environmental conditions."[7]

Obviously we would all like to know exactly what it is that determines whether or not a specific bout of viral infection might lead on to cancer, and whether or not anything can be done about it. Understanding in this area is most likely to follow on from an answer to the really central question of how can viruses cause cancer in any case? The search for the *molecular mechanisms* by which viruses cause cancer has without doubt been one of the most active areas of scientific research in recent years. It is a search that is not only revealing the details of viral carcinogenesis, but which also seems to have uncovered vital clues to the origin of all types of cancer, whether produced by chemical carcinogens, radiation or whatever. We shall be turning to these discoveries very soon; but before we can examine *how* the viruses might be able to cause cancer, we must first consider exactly what cancer really *is*.

What do they do?

It is one thing to identify a series of viruses that can cause cancer, but quite another thing to try to "look" inside the infected cells and find out what the viruses do to turn them into cancer cells, and even before examining the cancer cells, the first problem is to identify the results of cancer overall.

The most obvious feature of all cancers is that they arise from the uncontrolled multiplication of cells whose growth and multiplication is normally kept strictly under control. Any new and abnormal growth of cells is called a "tumour", but there are many types of tumours that are not cancers. Warts, moles and certain cysts are well-known examples of relatively harmless or "benign" tumours. They remain isolated and do not spread to other parts of the body or invade and destroy nearby tissue.

Cancers, on the other hand, are a specific class of tumours with various extremely damaging or "malignant" characteristics. They *can* invade and destroy nearby vital tissues of the body. They can spread throughout the body when individual cells are shed from the original tumour and carried away by the circulatory system to settle and develop into "secondary" cancers elsewhere (a process known overall as "metastasis"). Cancer cells lose the normal appearance and characteristics of specialised healthy cells, reverting to a more primitive or embryonic state which is accompanied by an endless cycle of cell multiplication. In contrast to benign tumours, malignant tumours are of course far from harmless. Without medical intervention most cancers (once they have escaped the body's anti-cancer immune defences and fully established themselves) lead on to inevitable death within a fairly short time.

The disease of cancer, then, involves the many millions of cells that comprise a malignant tumour and can be present at several distinct locations throughout the body. But an understanding of the disease overall is made simpler by the fact that cancer can begin as just one abnormal cell, which then produces the entire tumour mass by uncontrolled cell growth and division. This ability of just one cell to cause cancer was demonstrated many years ago by isolating single cancer cells, transplanting them into healthy animals, and observing that full-sized cancers could then develop. These sort of experiments have justified an attack on the problem of cancer causation by studying cultured cancer cells rather than live diseased animals.

Working with colonies of cultured cells under controlled laboratory conditions is of course much easier than trying to deal directly with intact animals. Almost all that is known about the ways in which viruses can cause cancer was initially discovered by using cultured cancer-like cells as models of the disease. In their excitement over the results that have emerged from such studies scientists have sometimes left themselves open to criticism for forgetting that individual cells that behave like cancers in a culture dish are not really cancers.[17] While the case for initially investigating the disease at the level of individual cells is a strong one, cancer is by definition a multi-cellular disease which interacts in many complex ways with the healthy organs, tissues and cells of its victims. Any results from cell culture studies should therefore be treated with a certain amount of caution.

Having summarised what cancer is, and having explained why scientists have decided to attack the problem of cancer causation right down at the level of individual cells, the next problem is to define the particular characteristics of a cancer cell that allow the cancer overall to behave in the manner outlined above. Cultured healthy cells can be converted into cancer-like cells by a process known as "transformation", which can be brought about by cancer viruses, chemical carcinogens or radiation. The overall effect of transformation is to shift a cell into a cycle of unregulated growth and multiplication very similar to the behaviour of real cancer cells. But what exactly is different about a transformed cell?

If normal cells are added to a culture dish, along with a supply of nutrients, they will grow and multiply until a flat sheet of cells covers the available space on the bottom of the dish. At this point the cells will stop dividing, further multiplication being held in check by what is known as "contact inhibition". The meaning of this term is fairly obvious: once normal cells come into contact with other cells all around them their growth is inhibited by the presence of, or direct contact with, these other cells. Cells of the multi-cellular higher organisms always have to live in harmony with their millions of fellow cells that make up the organism as a whole. Contact inhibition would therefore seem to be a very "sensible" way of ensuring that no single cell or cell type multiplies out of control to the detriment of all the other cells nearby.

In addition to contact inhibition, normal cells grown in culture exhibit a number of other properties that seem to characterise well-regulated "socially acceptable" cell behaviour (see figure 10.1). They can only grow if attached to some rigid surface such as the bottom of a culture dish – a phenomenon known as "anchorage dependence". They will stop growing if the supply of nutrients from the culture medium begins to run out. And they are not "invasive", meaning that they cannot invade and eventually destroy other types of multi-cellular tissue.

Transformation, for example by infection with a cancer virus, can free cells from all these social constraints. Growth will no longer be held in check by contact inhibition and so the transformed cells will multiply out of control, piling up on top of one another to form lumps in the otherwise flat sheet of normal cells. Anchorage dependence is lost, allowing transformed cells to grow and multiply while suspended in a nutrient solution; and they no longer stop growing if the nutrient supply begins to run out, continuing instead in a cycle of reckless multiplication until it kills them. Transformation also makes the cells invasive and so able to eat into foreign tissue in the characteristic fashion of a cancer cell. One additional freedom bestowed upon a transformed cell is freedom from the need to die. Most normal cells can establish cultures that will only last for a certain time (about 50 cell generations for human cells); but cultures of transformed cells become immortal, growing and dividing endlessly provided an adequate supply of nutrients is maintained.

All of these altered properties of transformed cells fit in with the observed uncontrolled and indefinite growth of a cancer cell. They are accompanied by specific biochemical changes within the cells that presumably mediate the overall effects of transformation just outlined. For example, new genes become active within the nucleus of a transformed cell and new proteins appear within the cytoplasm. The chemical composition of the cell surface changes, with some new proteins and carbohydrates appearing while others disappear. The rate at which some specific materials are transported into and out of the cell becomes altered. The task faced by

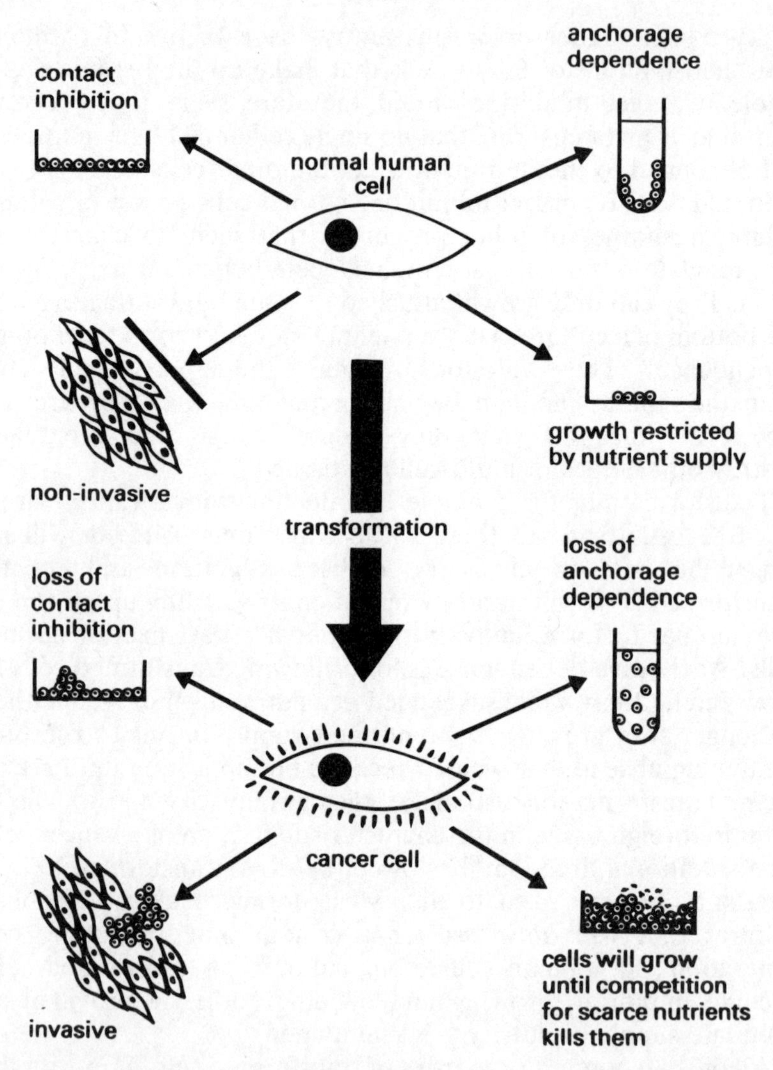

Figure 10.1 Some of the major changes brought about in cells when they are "transformed" into cancer cells

cancer researchers is to explain how carcinogens, including some viruses, can bring about such drastic changes in a previously healthy and well-regulated cell.

The process of understanding cancer has gradually focused in on the disease at successively increasing powers of magnification; first of all describing the behaviour of cancers overall, next looking in on the altered behaviour of the individual cells involved, and then peering right down to the specific biochemical changes that characterise a cancer cell. The final, most powerful lens of the analytical "microscope" has recently snapped into place and scientists throughout the world are busy adjusting the "focus", perfecting the "lighting", and at last discovering some of the key molecular steps that first send cells down the road to cancer. It is the study of how *viruses* can cause cancer that has provided the first clear enlightening glimpses.

How do they do it? – Integration, cancer genes, cancer proteins

Many different types of virus can produce cancer, both those with a DNA genome and those whose genetic material is made of RNA. But interestingly, all of the RNA-containing cancer viruses are *retroviruses*. The retroviruses, remember, make a DNA copy of their genome which then integrates into the chromosomes of the infected cell. That the retroviruses are the only RNA viruses known to cause cancer should assume increased significance when I tell you that most cells transformed into cancer cells by *DNA*-containing viruses also carry the viral genes in an integrated form. We have already seen, for example, that liver cancer cells often contain integrated hepatitis B virus DNA.

A general rule emerging from studies of viral carcinogenesis is that *when viruses transform cultured cells or produce real cancers the viral genes must usually become integrated into the cancer cell DNA.*[18] Of course by no means all of the viruses capable of transforming cells have been fully investigated, and there is evidence suggesting that in isolated cases integration is *not* an absolute requirement for carcinogenesis. But from the results so far we can certainly say that at the very least integration must make it much easier for a virus to cause cancer.

Why should integration be so important? At first sight there are at least two possible answers. One is that the actual act of integration is somehow responsible for producing the cancer. For example, the viral genes might integrate into the middle of, and therefore destroy or change, genes that are important for keeping the host cells healthy. Alternatively, simply by integrating nearby important cellular genes the viral genome might increase or decrease the activity of these genes, again leading to cancer. It is certainly well known that the activity of genes can be strongly influenced by nearby sections of DNA, so the idea that neighbouring viral genes might disturb normal gene activity makes good sense.

Other than the possibility that integration might directly induce the cancer, however, another equally valid possibility is that the change to a cancer cell could be induced by specific viral proteins. In this case integration might simply serve to ensure that the viral genes responsible for encoding such "cancer proteins" become a permanent part of the cancer cells' genetic material. After all, a cancer is a mass of rapidly dividing cells and if the viral genes did not become integrated then their effect might quickly be diluted. The speed of cell division might prevent sufficient viruses from being produced to keep all the new cells cancerous. Only integration of the viral genes would ensure that each new cell contained a copy of these genes to keep it cancerous.

The evidence emerging from the world's research laboratories suggests that all of the possibilities considered above might be relevant. Cancer may sometimes be produced by viral proteins, sometimes result from integration of viral DNA at a sensitive site, with integration in both circumstances serving to pass on the viral genes to every cell in a rapidly expanding cancer cell population. Many cancer viruses certainly *do* produce "cancer proteins" that are directly responsible for transforming healthy cells into cancerous ones. And of course cancer proteins must be encoded by genes – the now famous "cancer genes".

To pick up the threads of the cancer gene story we must first of all consider a very rare and unusual class of retroviruses known as the "acute transforming retroviruses". These viruses have been isolated from various animal cancers throughout the world, and when artificially administered to healthy animals they can produce cancer very quickly and with impressive regularity. One of them is the chicken cancer virus discovered by Peyton Rous in 1911, which will transform virtually every cell it infects. Unlike the acute transforming retroviruses, most other viruses linked to cancer can transform cells only infrequently, inducing the cancer-like state in only about one in every 1000 to 100,000 infected cells.

The most significant discovery about the acute transforming retroviruses is that they owe their cancer-causing prowess to specific genes that do not seem to have any other role to play in the life-cycle of the virus.[19] These cancer genes (or "oncogenes") were first detected by studying mutant viruses that behaved as if a mutation in some gene had removed their ability to cause cancer. As the modern techniques of molecular biology were developed it became possible to chop up the viral genetic material (in its double-stranded DNA form) into gene-sized pieces. These pieces were then artificially introduced into cultured cells, where they would sometimes integrate into the cell chromosomes and be used to make protein. This sort of experiment revealed that each acute transforming retrovirus carries one gene – the cancer gene – that can *on its own* transform cultured cells into cancer-like ones.

Just over 20 of these retroviral cancer genes are now known, having been isolated from different viruses that were found in chickens, turkeys, cats, rats, mice and monkeys. The discovery of single genes that can apparently cause cancer obviously aroused great excitement. Clearly the next step was to try to work out exactly what the proteins encoded by these cancer genes do. For most of the genes progress towards that goal is still rather limited; but in mid-1983 two separate research teams claimed that for the first time the likely function of a viral cancer protein had been identified.

The two teams (one led by Russell Doolittle of the University of California and the other by Michael Waterfield of the British Imperial Cancer Research Fund) had discovered that the cancer protein produced by a retrovirus found in monkeys was almost identical to a naturally occurring human "growth factor".[20,21] As the name suggests, growth factors are proteins that can stimulate cells to grow and divide. Now since cancer is essentially unregulated cell growth and division, it makes good sense that viruses should use growth factor, or growth factor-like, proteins to produce cancer.

The announcement of the similarity between a viral cancer protein and a natural growth factor was hailed as a major event not only by the scientific press, but also on television and in the mass circulation daily newspapers. As is almost always the case in modern science, the praises heaped upon the workers who made the final discovery should also have been directed at the large number of other scientists whose collective efforts had made it possible. Nevertheless, the excitement was certainly justified. For the first time a logical path was becoming evident leading from a carcinogen (the virus) all the way to the eventual unregulated growth of the cancer cells. This will undoubtedly come to be recognised as one of the major milestones on the way to a complete understanding of cancer.

More recent research has confirmed that the product of the viral cancer gene really does *act* very much like a growth factor,[22] and has also uncovered another retroviral cancer protein whose activities can be fitted into the same general cancer-causing scheme. In this second case the cancer protein is not a growth factor, but instead appears to resemble part of the cell-surface receptor that *binds* to extracellular growth factors and communicates the "start growing" message to the rest of the cell.[23] Again, the activity of this cancer protein within the cell presumably stimulates the normal cellular processes that lead to growth and division, this time by mimicking the next step on in the natural sequence (see figure 10.2).

The precise functions of the proteins encoded by the other retroviral cancer genes have not yet been revealed, although many people are busy investigating the problem. It seems highly likely that they will all be able to mimic the activity of cellular proteins involved in the normally rigidly controlled pathways leading to cell growth and division. The ability to produce proteins that "mimic" such crucial normal cell proteins might seem to be a clever trick that the acute transforming retroviruses use to produce an expanding population of virus-infected cancer cells. In fact this "trick" is not what it seems. The retroviral cancer genes are not the end products of a long process of viral evolution; in fact they are not really viral genes at all. Instead, they are altered copies of the crucial cellular genes themselves, "stolen" by retroviruses from the chromosomes of the cells they infect.

To cut an extremely long and complicated story short (since we are more concerned here with what the viruses *do* than in how we found out that they can do it), it appears that when retroviruses infect a cell they can sometimes pick up a cellular gene and incorporate it into the viral genome[24] (see box 10A). The activity of the cellular gene will then be controlled by the regulatory regions of the viral genome, which might make it much more active than it should be in a healthy cell. Alternatively, the alterations in the cellular gene that usually accompany the act of hijack might make it produce an altered protein, and this altered protein might produce cancer rather than cell growth only when required. I should point out that *any* cellular gene can probably be picked up by a retrovirus and incorporated into its genome, but only if the

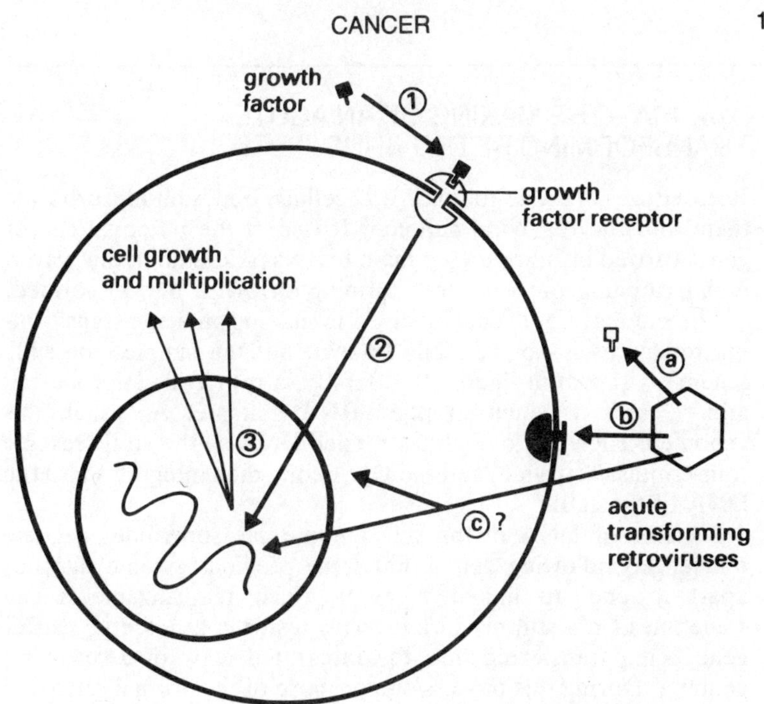

Figure 10.2 How acute transforming retroviruses cause cancer. In healthy cells growth factors can bind to a cell-surface receptor ①, causing the receptor to initiate cell growth. It probably does this by somehow activating various genes in the nucleus ② which then make the proteins that actually bring about growth ③. Each acute transforming retrovirus carries a cancer gene (actually a stolen and probably altered copy of a crucial cellular gene – see box 10A and text). One of these cancer genes apparently makes a growth factor or growth factor-like protein ⓐ, while another makes part of a growth factor receptor protein ⓑ. In either case inappropriate cell growth and multiplication will be stimulated within infected cells, leading to cancer. Other retrovirus cancer genes may code for proteins that can intervene at other points in the normal cell growth pathway ⓒ

gene is a crucial one involved in the control of cell growth will the resulting hybrid retrovirus be able to cause cancer.

So the acute transforming retroviruses are formed when a retrovirus picks up a normal cellular gene involved in cell growth and multiplication. The cell gene will either be crucially altered during this process, or else might be expressed at abnormal levels, in both cases leading to the onset of cancer (see figure 10.4a).

BOX 10A – THE MAKING OF AN ACUTE TRANSFORMING RETROVIRUS

Retroviruses are able to "pick up" cellular genes and incorporate them into the retroviral genome. If one of the potential cancer genes carried by all cells (see main text) is picked up in this way, a highly carcinogenic acute transforming retrovirus may be formed.

The sequence of events involved in making an acute transforming retrovirus is not yet fully known, but the simplest possible scheme is shown in figure 10.3. Firstly, a retrovirus infects a cell and releases its single-stranded RNA genome. As usual, this genome is copied into double-stranded DNA by the viral "reverse transcriptase" enzyme, and the DNA copy then integrates into the DNA of the cell.

Sections of DNA in the cell genome can sometimes become rearranged, allowing genes that were previously some distance apart to end up linked together. Such rearrangements can sometimes (in a still mysterious way) result in a potential cancer gene being transferred into the integrated copy of a retroviral genome. During this process one or more of the normal retroviral genes are usually lost or damaged (for example by insertion of the cellular gene into the middle of a retroviral gene). This damage explains why acute transforming retroviruses are usually "defective" – unable to multiply without assistance from normal retroviruses within the infected cell.

The potential cancer gene may also be altered during the rearrangement process, or it might simply end up next to regions of viral DNA that make it much more active than it is normally. Such changes in the gene's structure or activity are believed to convert it from a potential cancer gene into a cancer gene proper, encoding a cancer protein capable of transforming cells.

The rearranged genome can then be copied into RNA; and these RNA copies can be incorporated into protein coats to generate acute transforming retroviruses which will be released from the cell.

On entering another cell, the genome of an acute transforming retrovirus will be copied into DNA and integrated into the cell genome as usual, and once integrated, the retroviral cancer gene can begin making the cancer protein that makes the cell cancerous.

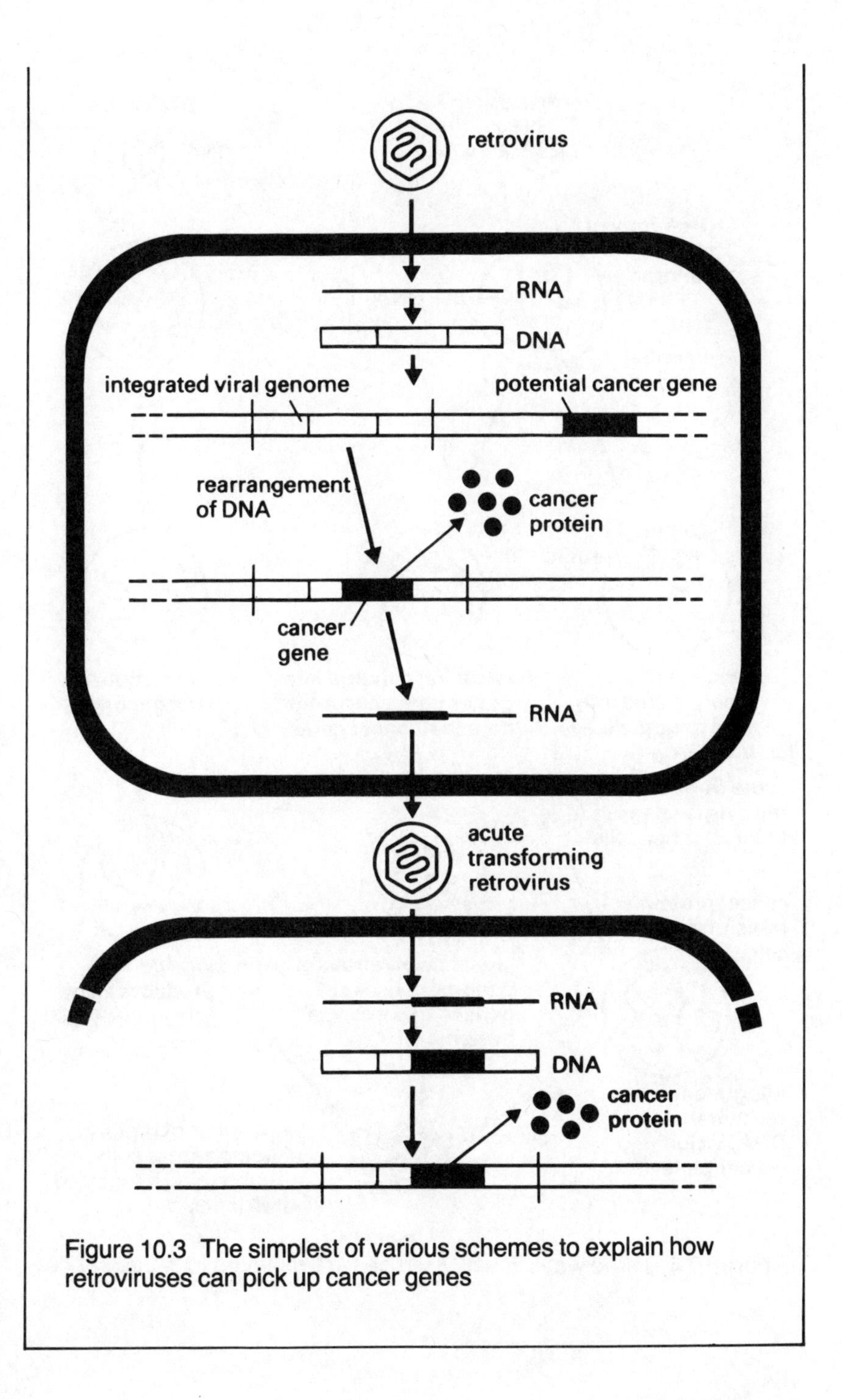

Figure 10.3 The simplest of various schemes to explain how retroviruses can pick up cancer genes

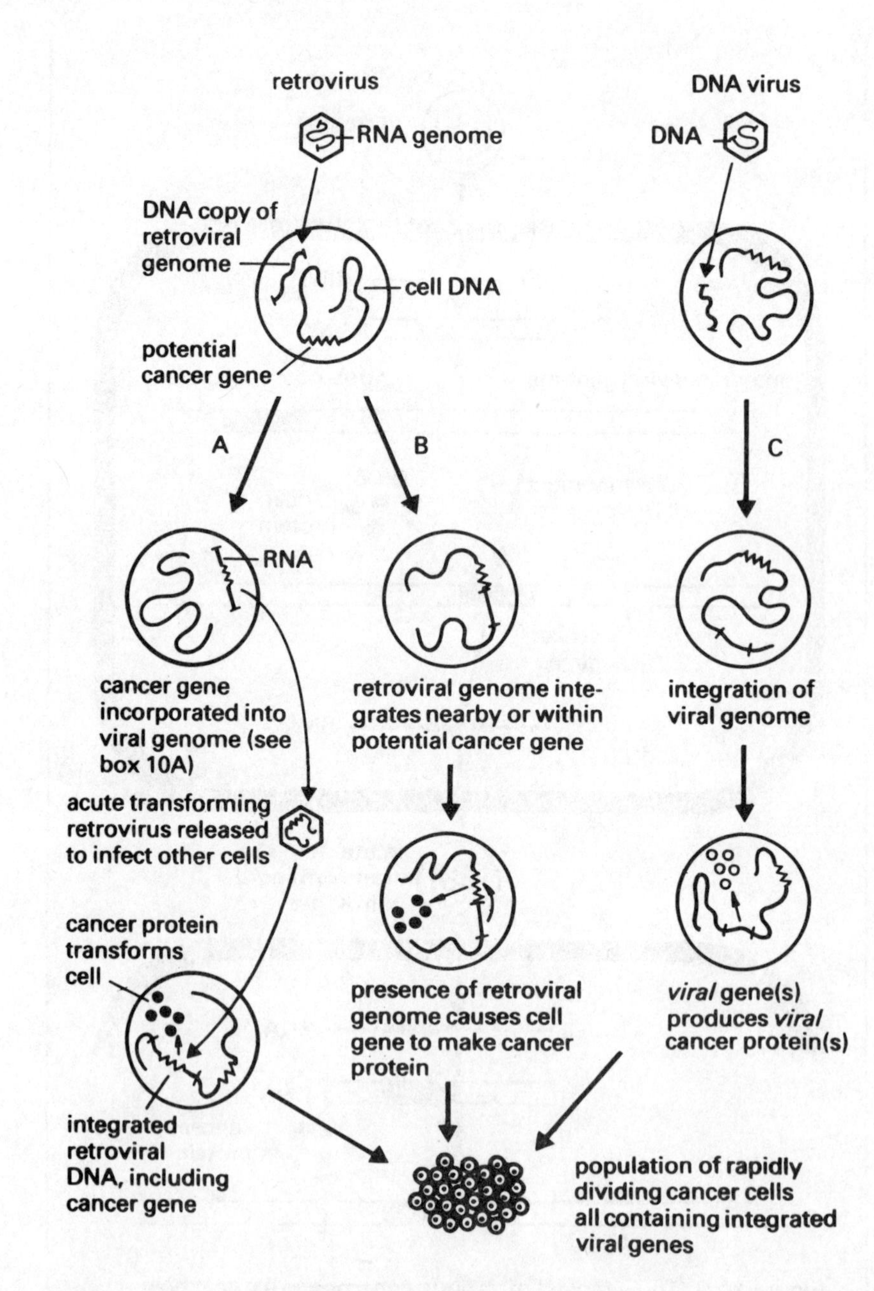

Figure 10.4 Three ways in which viruses are believed to cause cancer

Exactly what change in a gene's structure or expression can allow it to become a retroviral cancer gene is still the subject of speculation and further research. But the fact that retroviral cancer genes *are* hijacked from the cell in this way seems to be firmly established.

This tale about the formation of the acute transforming retroviruses explains some of their unusual features. One of the most unusual things about them is that they do not seem to be transmitted naturally between animals, either by infection or from parents to their offspring. There are no reports of them ever having caused epidemics of cancer, and we only know of their existence because they have been isolated from a few spontaneous cancers found in various types of animal. They appear to arise within an individual animal (when an infecting retrovirus picks up one of the crucial genes), produce cancer, and then die along with their infected host. Most of them are actually *defective*, lacking one or more of the genes needed for them to complete the normal retroviral life-cycle. They can multiply successfully, however, when normal viruses infect the cell along with them and perform the tasks that the acute transforming retroviruses cannot do for themselves. The reason why they are often defective is that when the cellular gene is added into the middle of the retroviral genome, it can cause the loss of or damage to some vital genes of the retrovirus itself.

Despite the rarity of the acute transforming retroviruses, and their many unusual characteristics, scientists know more about how they can cause cancer than they do about any other type of virus. At first sight this might seem to be highly unfortunate, since the way in which these unusual viruses cause cancer might have told us little about other more common forms of carcinogenesis. In fact, the opposite appears to be true, with the acute transforming retroviruses pointing the way towards a full understanding of cancer in general.

Cast your mind back to the two main alternative explanations offered for the importance of integration in viral carcinogenesis. The first suggestion was that by integrating into, or near, critical cellular genes a virus might alter the activity of these genes and so give rise to cancer. The study of the acute transforming retroviruses has revealed that there are at least 20 or so suitable genes within normal cellular DNA. The acute transforming retroviruses alter the activity of these genes by actually having them incorporated into the viral genome. Might not other viruses alter the activity of these genes simply by integrating into or next to them? There is a class of cancer-causing retroviruses known as the *slow* transforming retroviruses that may well be able to transform infected cells and cause cancer in just this way.[25]

The slow transforming retroviruses do not pick up the cellular genes that appear to have the potential to cause cancer, but in some cancers the viral genomes have been found integrated next to or actually within these crucial genes. So the theory about these retroviruses is that they cause cancer by "switching on" the activity of genes involved in cell growth and multiplication simply by integrating near to them on the host cell chromosome; or by integrating *into* a gene they might make the gene code for an altered protein with an aggressive cancer-causing capability (see figure 10.4B). As their name suggests, cancer brought about by the slow transforming retroviruses is a slower and less inevitable process than carcinogenesis due to acute transforming retroviruses. An acute transforming retrovirus comes supplied with its own cancer gene, so the transformation of an infected cell may be almost inevitable. For a slow transforming retrovirus to cause cancer it must first integrate next to an appropriate gene in an appropriate way – events which might both be rather unlikely. In rare cases the very viruses that normally behave as slow transforming retroviruses might be able to pick up one of the crucial cellular genes and so be converted into acute transforming retroviruses.[26]

So retroviruses in general have at least two paths towards causing cancer available when they infect a cell (the two diverging paths of figure 10.4 A and B). They can incorporate genes involved in cell growth into the viral genome; changing the virus into an acute transforming retrovirus that will spread throughout the host cells, cause cancer, and probably die along with the host. Or they can influence the important cellular genes simply by integrating within them or nearby. Most retroviral infections of course, probably follow neither path – most of them are harmless.

So to summarise where we have got to so far; it appears that normal healthy cells carry genes that, if abnormally expressed or altered in some way, have the potential to produce proteins that can cause cancer. When these genes become incorporated into the genome of an acute transforming retrovirus, or when they are activated by the integration of a slow transforming retrovirus, then we can justifiably call them "cancer genes" – since in such cases they clearly do cause cancer. But when the genes are sitting in their normal place within the cell chromosomes, and are held in check by the control systems of a healthy cell, then "*potential* cancer genes" is their most appropriate label ("proto-oncogenes" is the technical term). This label acknowledges the fact that these genes *can* cause cancer if they run out of control or are altered in some way, while recognising that normally they do not cause cancer at all. Indeed they are believed to play absolutely crucial roles in the life-cycle of healthy cells – presumably mediating many of the key events during normal cell growth and multiplication (by encoding growth factors or growth factor receptors, for example).

Similarly the proteins produced by the potential cancer genes only become "cancer proteins" when they are present in excess quantities or altered in some way, in both cases leading to cancer.

This discussion of the potential cancer genes sitting within all our cells brings us to the much wider and more significant subject of carcinogenesis in general. Cancer is of course caused by many factors other than viruses. Possibly far more important than viruses are the chemical carcinogens, radiations and so on that are the traditionally accepted causes of cancer. Do the potential cancer genes revealed by studies of how retroviruses cause cancer have any role to play in non-viral carcinogenesis? The emerging evidence suggests that they may. Various different experimental approaches are suggesting that in many cases (but not always) the cancer genes at work in many non-viral cancers may be the *same* genes as have been shown to be involved in retroviral carcinogenesis. *So both retroviruses and non-viral carcinogens may well cause cancer by converting one or more members of a small set of crucial cellular genes, the potential cancer genes, into cancer genes proper.* Current estimates suggest that there are probably less than 50 of these potential cancer genes available (and perhaps as few as 25 to 30).[27]

An appealing simplicity and unity, then, appears to be emerging from research into both viral and non-viral carcinogenesis: The healthy cells of the body contain the potential "seeds of their own destruction" in the form of a small set of genes that control cell growth and multiplication when working properly, but which can cause cancer when they become altered or forced to run out of control (see figure 10.5).

But there is an extremely important viral piece of the cancer jigsaw that has yet to be put in its place – the involvement of the *DNA*-containing viruses in causing cancer. Can the activities of

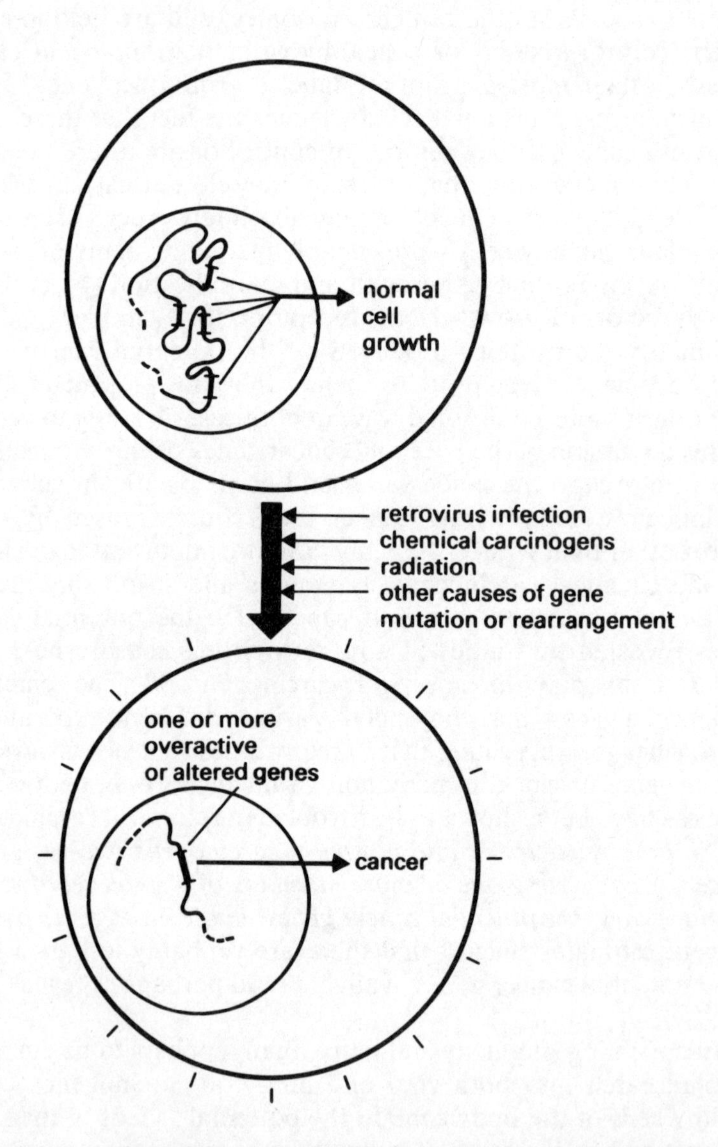

Figure 10.5 A unified theory of retroviral and non-viral carcinogenesis. Healthy cells contain a set of genes that normally mediate controlled cell growth and multiplication. Retroviruses and other carcinogens cause cancer by altering these potential cancer genes, or by making them overactive

viruses that carry a DNA genome be fitted into the neat overall cancer gene pattern of figure 10.5?

A first look at the evidence would suggest that the answer must be no. The DNA viruses do not appear to pick up potential cancer genes from cellular DNA. Some of them *might* activate these cellular genes by integrating nearby, but that has yet to be proven. What *is* known is that some DNA viruses carry truly viral cancer genes, which are permanent parts of the viral genome and which make viral proteins that are able to make an infected cell cancerous (see figure 10.4C). Studies of the cancer genes carried by DNA viruses, and their associated cancer proteins, have lagged a bit behind research into retroviral carcinogenesis; but some recent evidence suggests that the cancer genes of some DNA viruses produce proteins that perform very similar tasks to the proteins encoded by retroviral and cellular cancer genes.[27] So it may well turn out that at least some DNA viruses cause cancer by making viral proteins that really do "mimic" the activity of the now familiar set of potentially carcinogenic cell proteins.

Amidst all this detail about how various retroviruses and DNA viruses might cause cancer, I should point out that the way in which any specific viruses cause *human* cancers remains unclear. At the moment we can only *assume* that viruses such as hepatitis B, Epstein–Barr and so on cause cancer via the sort of mechanisms being revealed by studies of animal cancers and transformed cultured cells. The retrovirus HTLV may actually cause cancer by a quite different mechanism than those considered above. It does not seem to carry a cancer gene, but there is some evidence to suggest that it makes a viral protein which may be able to *switch on* cellular cancer genes. If further research confirms this model, then HTLV may have revealed a third major way in which retroviruses can cause cancer.

A great deal remains to be discovered about both virally induced cancers and all other types of cancer, but new results are now flowing thick and fast. After many years of stubborn resistance the barriers separating us from a detailed insight into how cancer cells arise and work do seem to be coming down at last. At the time of writing it is becoming apparent that the activity of just one cancer gene might be insufficient to produce a real cancer (rather than just transform a cultured cell). In at least some cases the combined effects of at least two and perhaps three or more cancer genes might be required.[27] Other research is revealing a battery of "cancer suppressor" genes carried by healthy cells, whose normal job is to *stop* cancer from developing, and it seems that some cancers might arise when a cancer suppressor gene *stops* working, rather than simply when a cancer-causing cancer gene *starts* working.[28]

There will doubtless be many more fascinating and hopefully enlightening developments in the story of viral and non-viral cancers throughout the next few years. The neat overall schemes of cancer causation outlined in figures 10.4 and 10.5 might well come to be seen as gross over-simplifications. The only thing that can be said with any certainty at the moment is that there is a lot more work to be done before the mystery of cancer is finally solved, and there are probably many more surprises in store. No matter how interesting all the recent discoveries about cancer genes have been, they have not yet brought us to the point where we can start to rationally design much more effective therapies to combat cancer.

CHAPTER 11

Vaccines – priming the trap

In the entire history of medicine there can be few more significant achievements than the development of vaccination. The simple procedure of briefly exposing people to all or part of the infectious agent you wish to protect them against has given us the ability to prevent such important viral diseases as smallpox, yellow fever, rabies, polio, measles and mumps. Of course vaccination is not restricted to viral diseases. It has also been highly effective against bacterial infections such as diphtheria, tetanus, cholera, whooping cough and so on. But it is probably against the viruses that the results of vaccination have been most beneficial and dramatic, particularly considering the long-standing dearth of any really effective anti-viral drugs.

Smallpox, for example, has been completely eradicated thanks to a global vaccination programme; and many other viral infections might eventually meet the same fate if the necessary money and political will are forthcoming. But the story of vaccination is not entirely a tale of success. Some important viral infections have resisted the technique – due to some troublesome properties of the viruses concerned, the deficiences of vaccine technology, or both. Hepatitis, herpes and influenza are just three groups of viruses that have for a long time been chapters of failure, or only partial success, in the otherwise cheerful vaccination story. But the 1980s are unfolding as an exciting and revolutionary new chapter, as developments in biotechnology promise to yield effective new vaccines for many previously intractable viral diseases.

Many people will tell you that vaccination began in 1798 with Edward Jenner, an English doctor with a country practice in

Gloucestershire where the first really safe vaccines were developed. We shall return to Jenner's contribution shortly, but the theory that prior exposure to a small amount of infectious agent can subsequently protect against the full-blown disease has been practised for thousands of years. Many ancient civilisations knew that babies and young children could be protected against smallpox by introducing into the bloodstream a small amount of infected material taken from a victim. This practice was certainly very common throughout Europe in the years preceding Jenner's celebrated experiments; and it was apparently quite successful, with relatively few mishaps. The likely "mishap", of course, is that too much or too virulent a preparation of infected material might be used, resulting in death from the very disease it was hoped to prevent.

Jenner's great contribution was to send the ancient procedure down the road towards *safe* vaccines that could be used with little risk of the recipient becoming a victim. His interest was aroused by the observation that infection with cowpox (a similar disease to smallpox, but milder and not fatal to humans) appeared to bestow subsequent protection against the more serious smallpox. Cowpox was commonly contracted by people in close contact with cows, such as milkmaids, farmhands and so on. Jenner gained personal experience of the protection offered by cowpox while performing the ancient procedure of inoculating people with smallpox-infected material in order to protect them from the disease. He found it impossible to produce the mild illness associated with a successful inoculation in a man who had previously caught cowpox.

Jenner decided to test the potential of cowpox protection by artificially infecting a healthy boy with matter from a cowpox-infected sore on a milkmaid's hand. The boy developed the usual mild symptoms of cowpox and then recovered. About 6 weeks later Jenner inoculated the boy with smallpox. Fortunately for the boy, Jenner and medicine in general, no illness was produced.[1] So Jenner had confirmed that you could *safely* protect someone from a dangerous viral disease by infecting them with a *related but less dangerous* type of virus. The significance of Jenner's work is immortalised in the very term "vaccination" (vaccinia = cow), which now applies to "vaccines" used against many other infections that have nothing to do with cows or cow viruses.

Of course when Jenner developed his smallpox vaccine he did so in complete ignorance of both the nature of the infectious agent causing the disease and the way in which his technique worked. His research was a triumph for the sort of experimental approach that even today allows us to produce effective therapies for diseases that are not yet fully understood. Let's now leave the mysterious world of Edward Jenner and jump back to the present, looking at the practice, problems and future potential of vaccination in the light of modern knowledge about viruses and how the body's immune defences work.

Strategies

Armed with the information given in chapter 7, the way in which vaccination works should be easy to understand. We saw in that chapter how the immune system can "remember" a previous infection, allowing it to quickly overcome any subsequent infection caused by the same organism. The "memory", of course, simply takes the form of an expanded population of B- and T-cells ready to recognise the organism and initiate the amplified immune response found in an "immunised" individual. The trick of safe vaccination is to expose the body to viruses (or other micro-organisms) in a form that triggers the immune system without causing any serious disease.

The ancient practice of using small amounts of a living virulent virus as the necessary trigger walked a chancy tightrope on which loss of balance by using too much virus could result in death. Jenner's safe vaccine exploited the fact that an immune response against the relatively harmless cowpox virus will produce an immune memory that is also effective against smallpox virus. For Jenner's procedure to work, some crucial antigens carried by the cowpox virus obviously had to be identical or very similar to antigens on the smallpox virus – so similar, in fact, that the cells of the immune memory could not really tell the difference between the viruses. With viral *antigens* we have arrived at the crucial determinants of the success of any vaccine.

Immunity, remember, is mediated by receptors on T- and B-cells that can bind to appropriate antigens (usually proteins or glycoproteins) on the surface of a virus. It is the viral antigens that are important, not the whole virus. The challenge facing the vaccine designer is to produce a preparation of suitable viral antigens that can stimulate an immune response without causing either a serious infection or any other damaging side-effects.

There are several fairly obvious ways to attempt to achieve the vaccine designer's aim (see figure 11.1). Firstly, you could use viruses related to the target virus that stimulate the immune system but do not cause serious disease (Jenner's approach). Modern anti-smallpox vaccines are also of this type, consisting of a living

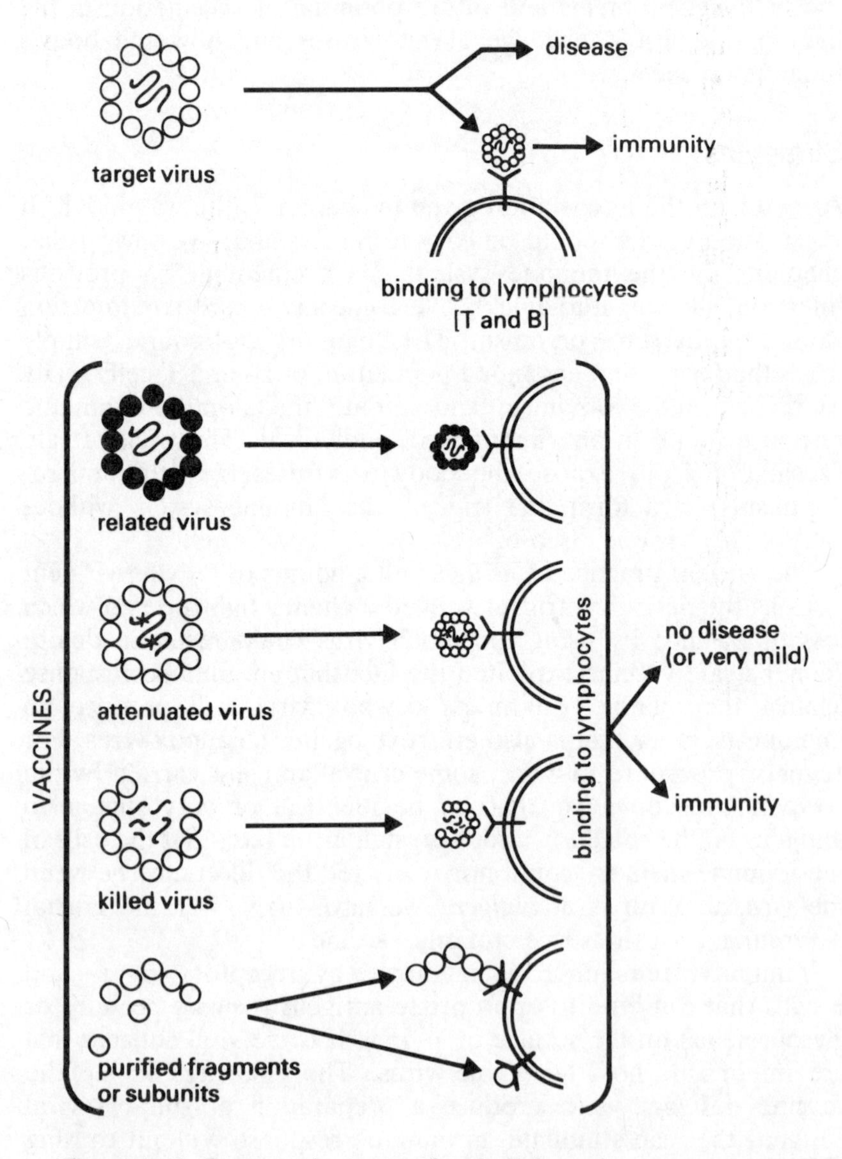

Figure 11.1 Some major types of anti-viral vaccines

virus (called "vaccinia" virus) that is derived from smallpox or cowpox virus and may be a hybrid between the two. Secondly, you could artificially create modified or mutant viruses that have lost their ability to cause serious illness but which still carry the antigens needed to stimulate the immune system. Such "attenuated" (weakened) live viruses are used in the modern vaccines against measles, polio, mumps and German measles. Producing a suitably attenuated form of a virus is traditionally a rather imprecise, fortuitous process. It has been discovered that when viruses are propagated in the artificial conditions of cell cultures they gradually become specialised to multiply in these cultures and become less proficient at multiplying within the body (a process that presumably involves the generation of new mutant viruses). So to produce an attentuated virus you just grow the original virus in some artificial culture conditions, and keep testing to see if it has changed into a form suitable for a vaccine. The virus used for most polio vaccines was produced by growing polio virus in cultured monkey kidney cells. For mumps vaccine the mumps virus was grown in chicken embryo cells, and so on.

A third approach to vaccine production is to "kill" the virus concerned in some way before using it as a vaccine. Remember that only the viral antigens are needed for the vaccine. So if a virus is "killed" (for example by using ultra-violet light or chemicals to wreck its genetic material and destroy its overall integrity) then the necessary antigens may remain intact while the virus will obviously be unable to multiply or cause disease. Anti-rabies vaccine is prepared by killing rabies virus with a chemical called beta-propiolactone. Killed viruses are usually less efficient than live virus vaccines since the viruses obviously cannot multiply at all after administration, so they generally induce a weaker immune response. For this reason killed virus vaccines often need to be given at regular intervals to maintain effective immunity. A single dose of some live vaccines, on the other hand, can give protection for many years or even a lifetime.

Finally, taking the principle that it is only the viral antigens that really matter to its ultimate conclusion, it is sometimes possible to make effective vaccines out of fragments of a virus or even the purified viral proteins. Obviously the safest type of vaccine would contain no viral genetic material whatsoever, ensuring that no unwanted infection could possibly arise. Some anti-influenza vaccines consist of fragments of the virus's protein coat that have been extracted from the virus into the solvent ether. We will meet such protein-only vaccines again when we come to consider the most recent developments in vaccine technology.

Regardless of the methods used to produce a vaccine, the sought-for properties are all the same. Obviously the vaccine must be *effective* at stimulating the immune system, producing an immunity that is as *long-lived* as possible. There should be no unacceptable *side-effects* of the vaccine; and if the vaccine is going to be of benefit world-wide then it must be *cheap* and *stable* for long periods of time (preferably without the need for refrigeration). These last two requirements are particularly important if the vaccine is to be successfully used in the Third World; where money is scarce, temperatures often high and refrigerators few and far between.

The vaccines commonly in use today are by no means perfect, but they usually do fulfil many of the ideal requirements given above. To gain an insight into the potential benefits of effective vaccines we need only consider the remarkable tale of smallpox. For millennia smallpox has been one of the greatest infective scourges of mankind, entering the body via the respiratory system, then multiplying and spreading by way of lymph and blood to cause high fever and death in up to half of all its victims. Even those who survived were usually severely disfigured by horrific scarring of the face, sometimes accompanied by blindness. Throughout the early 1960s over 15 million people a year were falling victim to this terrible virus, and yet on May 8th 1980 the World Health Organisation was able to triumphantly declare that it had been conquered. Not just controlled but completely eradicated; hopefully never to return.

Smallpox

The story of mankind's victory over the smallpox virus is an inspiring tale of global-cooperation towards a single goal.[2,3] It began in 1959 when the World Health Organisation (WHO) decided that a world-wide effort to eradicate smallpox virus was feasible and should be undertaken. The strategy was to be one of mass vaccination, accompanied by the identification of cases as soon as they occurred, isolation of the victims and vaccination of all their contacts. In this way it was hoped to deprive the virus of any susceptible hosts in whom it could multiply. WHO officials were encouraged to believe that global eradication was possible by the previous success of national eradication programmes undertaken by the richer nations. By 1959, for example, smallpox had already been virtually eliminated from the whole of Europe.

For the next 7 years the WHO's ambitious programme proceeded throughout the target areas (particularly Africa, South-East Asia and Brazil), but the results were disappointing. The hoped-for dramatic decline in smallpox cases did not occur. This initial failure, despite the vaccination of millions of people, could well have finished off the project altogether. But fortunately the WHO persevered and in 1966 resolved to step up and re-organise its efforts. About 5 per cent of the organisation's budget (of around $50 million) was committed to the smallpox programme and a much more reliable reporting system was set up to identify and then eliminate outbreaks of the disease.

From 1966 onwards the entire Third World was scoured literally village by village in search of the dreaded virus. In India, for example, over 100,000 health workers set aside 1 week per month for the smallpox search, and their progress was aided by the offer of cash rewards for the discovery of new cases. I should add that financial reward was probably also a major incentive encouraging the richer countries to underwrite the costs of the programme. Prior to eradication it was apparently costing millions per year to keep the richer nations free from smallpox by vaccination, maintaining quarantine barriers and so on – money that would be saved if the programme succeeded.

And from 1967 onwards it certainly *did* succeed. There was a sharp decline both in individual cases and the number of countries afflicted by the virus. The last-ever smallpox case in Brazil was registered in 1971. By 1975 Asia was smallpox-free, and then on October 26th 1977 hopefully the last-ever naturally occurring case of smallpox was reported to the WHO. On 25th July 1978, however, a medical photographer contracted the disease by her contact with a Birmingham University research laboratory in which the smallpox virus was being studied. This victim eventually died, some time after the scientist in charge of the laboratory had taken the blame for the incident and killed himself. This tragic episode highlighted the dangers of the stocks of smallpox virus held at a small number of research centres throughout the world, and brought about calls for these stocks to be destroyed. But provided such accidental infections can be avoided in the future, and provided governments do not turn to the smallpox virus as an instrument of war, then mankind will for the first time have completely conquered an infectious disease of major importance.

Obviously the stunning success of the battle against smallpox not only testifies to the value of the WHO, but also points the way forward to future battles against other viral infections. Polio, measles and mumps might be suitable targets, and the WHO have initiated the field trials of vaccines against hepatitis B virus (and therefore hopefully much liver cancer) already mentioned. But any optimism about future eradication programmes should be tempered by the realisation that in many ways smallpox virus was an *ideal* candidate for eradication – other viruses might not be so "easily" defeated.

In what way was smallpox virus an ideal candidate? Well first of all it caused an acute and dramatic illness which could be easily identified, not only by trained medical workers but also by uneducated villagers. This made it much easier to isolate and try to contain new outbreaks than would be possible with a less obvious infection. Next, victims did not become able to pass the disease on until the characteristic rash had begun to appear, so there was no long period during which an unidentified victim could unwittingly infect large numbers of other people. Perhaps most importantly, the virus did not *persist* in any of the victims who recovered from the acute disease, so the problem of persistent carriers (found for example with hepatitis B) did not arise. Inapparent infections were fortunately rare, again reducing the possibilities for unnoticed spread of the disease. And there was no major non-human animal "reservoir" available for the virus to multiply within. Many viruses, such as influenza and rabies, naturally infect not only humans but also other animals with which we regularly come into contact. With such viruses vaccination programmes would also need to be directed against the animals concerned – a formidable proposition. Finally, only one form of the smallpox virus existed (at least as far as the immune system was concerned) so a single type of vaccine was sufficient, and a very stable and efficient vaccine was available. The preparation of dried but still "live" virus that was used as a vaccine could remain potent without refrigeration for at least a month; and a health worker's supply for a week could fit into a shirt pocket. All these different favourable features of both smallpox virus and the available vaccine were undoubtedly a great help in taking the vaccination programme to the most remote and inhospitable parts of the world.

Most other important viral infections do not demonstrate such a happy combination of favourable features to assist in future eradication efforts. Hepatitis B infection, for example, can persist, often goes unnoticed and has provided a major challenge to the vaccine designers. Many other infections are also often inapparent, or perhaps infectious before symptoms develop. Influenza viruses keep changing, as we saw earlier, and can multiply in at least one animal reservoir (the duck); and at least 100 different viruses cause the common cold and it hardly seems feasible to vaccinate ourselves against them all. Despite such problems several other viral diseases may well be eradicated over the coming decades, with yellow fever, polio and mumps as three of the most likely and worthwhile targets. Without doubt the efforts to consign more viruses to the WHO's dustbin will be assisted by new developments in "biotechnology" – the detailed manipulation of biological systems on an industrial scale.

There are many targets for the biotechnologist interested in vaccine production to aim at. Hepatitis B, herpes and influenza have been cited as examples of important types of virus that have proved difficult to vaccinate against; while even some of the quite successful existing anti-viral vaccines could be improved in various ways. Few current vaccines meet all of the ideal requirements of life-long efficacy, cheapness, stability and freedom from side-effects.

Unwanted side-effects are probably the most publicised deficiencies of some modern vaccines. Obviously, if a virus used for vaccination is insufficiently attenuated or incompletely inactivated then in rare cases the vaccine may cause disease instead of preventing it. In the past, incompletely killed poliovirus vaccines (not the attenutated live vaccines more common today) have produced paralytic polio in large numbers of children. Another problem might be contamination of a vaccine with unrelated micro-organisms. During the war, for example, thousands of American servicemen were infected with hepatitis B virus carried in a contaminated yellow fever vaccine. Some vaccines can also induce serious allergic and autoimmune responses and many cannot be given in pregnancy (for fear of damaging the relatively unprotected fetus) or in illness. Any illness that diminishes the potency of the immune system, for example, might allow a normally safe live virus vaccine to multiply and cause disease.

So there are considerable incentives not only to produce vaccines effective against previously resistant diseases, but also to improve the efficacy and safety, and reduce the cost, of existing vaccines. The traditional approaches to vaccine design outlined in figure 11.1 will probably continue to play an important part in the development of future vaccines, but they will be increasingly supplemented by some of biotechnology's powerful new techniques.

Antigens unlimited

Everyone interested in science must be aware that the past decade has seen a revolution in mankind's ability to produce large amounts of natural protein molecules. This is the most celebrated achievement of various new techniques in molecular genetics that have been collectively dubbed "genetic engineering".[4] Genetic engineers can now extract a particular gene from one type of organism and insert it into the genome of a quite different organism. The gene for a desired human protein, for example, can be put into bacterial DNA. The "engineered" bacteria can then be easily grown in large quantities – all the time producing the wanted protein within the bacterial cells. The protein can then be isolated from the bacteria and put to work. Such medically important proteins as insulin, interferon and growth hormone are now being cheaply manufactured in bulk in this way. Genetic engineering is allowing rare proteins to be manufactured in quantities undreamt of during earlier times, when laborious purification from human or animal tissue was the only source of supply.

Viral antigens are usually proteins too, so the advent of genetic engineering has made it possible to put the genes for specific viral antigens into bacteria, yeasts, or cultured cells; and then grow up the recipient cells to produce cheap and very pure antigen preparations for use in new vaccines. Genetic engineering's first important success in vaccine production is likely to be the development of cheap, safe and effective vaccines to combat hepatitis B.

The drug company Merck, Sharp and Dohme, and the young biotechnology outfit Biogen, have both produced experimental anti-hepatitis B vaccines composed of a pure viral coat protein ("surface antigen") manufactured in genetically engineered yeast cells (see figure 11.2).[5,6] These vaccines have been tested in

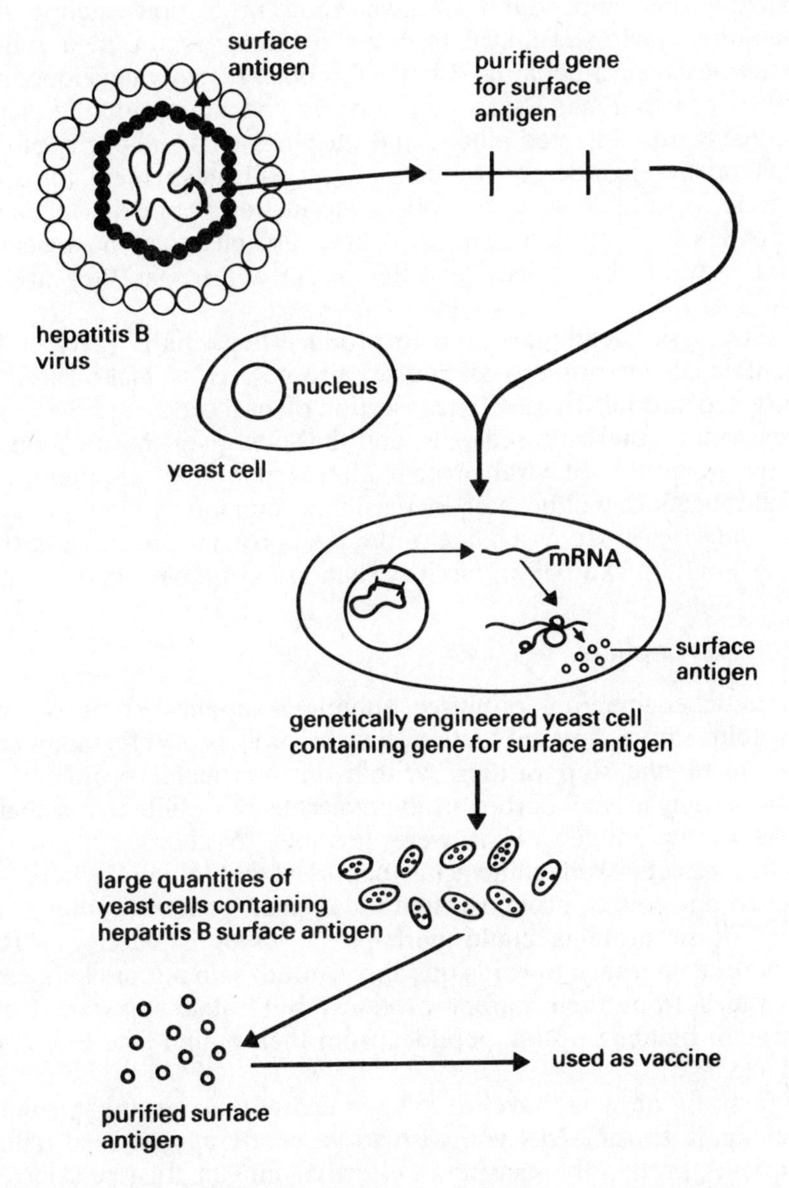

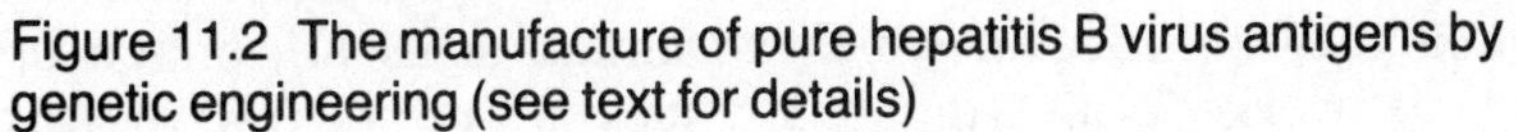

Figure 11.2 The manufacture of pure hepatitis B virus antigens by genetic engineering (see text for details)

chimpanzees and found to give good protection against the hepatitis B virus. Clinical trials on humans are the next stage. Vaccines against hepatitis B have also been recently developed by more conventional techniques such as the purification of viral antigens from infected blood;[7] but the cheapness and purity of the genetically engineered products might well make them the first vaccines suitable for a global vaccination campaign against hepatitis B. Such optimism, of course, assumes that the vaccines will prove to be at least as effective in humans as they are in chimpanzees.

The basic techniques used to produce hepatitis B vaccines by genetic engineering can, of course, also be used to make vaccines targeted at other viruses. The insertion of viral genes into bacteria, yeasts and other cultured cells, and the subsequent purification of large quantities of viral protein after a period of recipient cell multiplication, will become increasingly common through the mid and late 1980s. At least some of the viral proteins produced in this way should make better vaccines than are available today.

Building peptides

Genetic engineering promises unlimited supplies of pure viral proteins for vaccine production, but the process of refinement can be taken one step further. Within any particular protein it is usually only a small portion of the molecule as a whole that actually acts as an antigen. If it were possible to identify the short sequences of linked amino acids (peptides) that fold up to form the actual antigens of proteins, then these small peptides rather than the entire proteins could perhaps be used as vaccines. The favoured approach towards this aim is not to snip out the antigenic peptides from their parent proteins; but rather to start from scratch, building up the peptides from their amino acid building-blocks.[8,9]

First of all you have to take a close look at the proteins belonging to the virus you wish to vaccinate against (see figure 11.3). A protein that acts as an effective antigen must be selected and then the sequence in which its amino acids are linked up should be worked out (the techniques to do this were first developed in the 1950s and have now become routine). Having deduced the amino acid sequence of the protein as a whole, you would then manufacture peptides perhaps 10 to 50 amino acids in

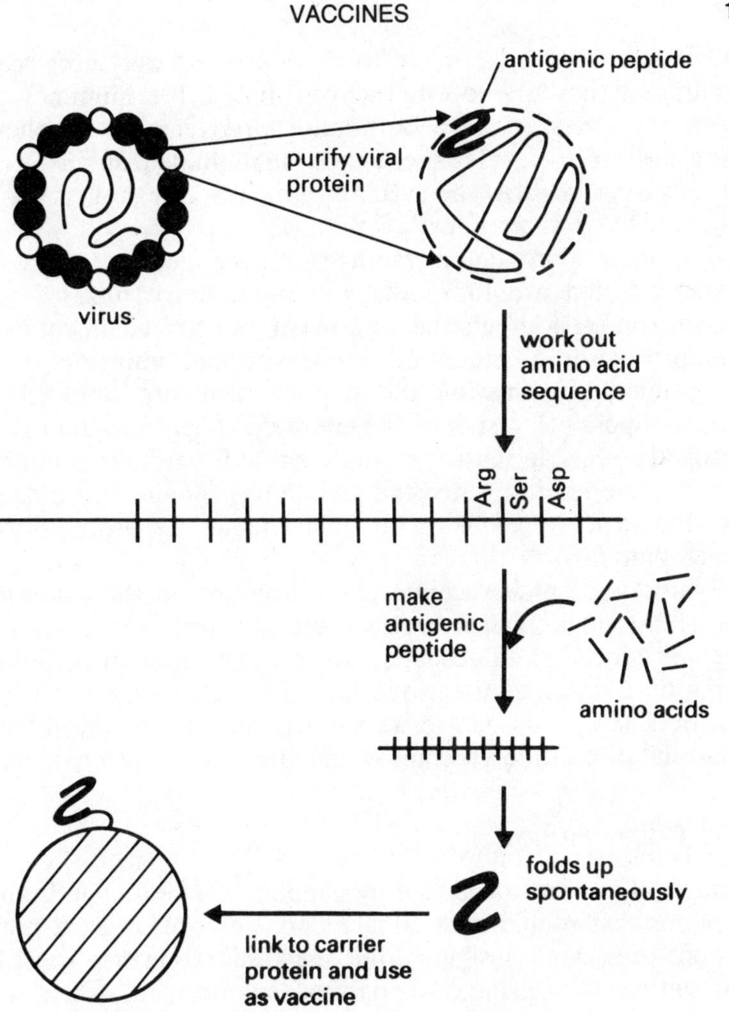

Figure 11.3 The manufacture of a peptide vaccine

length that match different regions of the viral protein. The chemistry required for such peptide manufacture is again becoming fairly routine. Hopefully one or more of the synthetic peptides will fold up to form a potent antigen, mimicking an antigenic site present on the protein. The potential of the various peptides could be tested by injecting them into animals to find out which ones stimulate an immune response effective against the virus as a whole. Any successful peptides may then be used as vaccines.

In practice a combination of different peptides might be best, and they will probably need to be linked to large inert "carrier" proteins if they are to efficiently stimulate the immune system. Also, they may need to be administered along with chemicals known as "adjuvants" that enhance the immune response, perhaps by ensuring a slow and steady release of antigen from an adjuvant/antigen complex.

The main potential advantages of the peptide approach to vaccine design are low costs and the relative ease of vaccine production on a large scale. It also offers precise control over the immune reponse generated by a vaccine, allowing it to be "fine-tuned" by varying the peptide structure until the most effective possible response is generated. Such fine-tuning is also becoming feasible with the whole protein vaccines produced by genetic engineering, since the techniques required to tinker with the fine structure of the genes that encode the proteins are also developing fast.

Synthetic peptide vaccines are still largely at the experimental stage. Vaccines against foot-and-mouth disease in animals have been produced, and research towards anti-influenza, polio and hepatitis B vaccines for use in humans is currently under way. It will be a few years before we see whether or not the theoretical potential of peptide vaccines is actually fulfilled in practice.

Rebuilding viruses

Pure viral proteins produced by genetic engineering and synthetic peptides that mimic viral antigens are perhaps of great potential benefit to vaccine designers, but they fall a long way short of the elusive "ideal" vaccine. One particular deficiency is their inability to multiply after administration, which might make repeated vaccinations necessary throughout a person's life. The ultimate answer to the vaccine designer's dreams would be the ability to construct novel live viruses which could be tailor-made to stimulate immunity without any risk of disease. Various research groups have recently taken the first steps towards this goal, not by constructing viruses anew, but by rebuilding existing viruses into forms that are more suitable for use as vaccines.

Once again it is genetic engineering that is clearing the way. Many different complex and versatile techniques are covered by the blanket term "genetic engineering". It would be inappropriate to go into all the details here but I can certainly summarise the things that genetic engineering makes possible. Very simply, it is gradually giving us mastery over the genomes of all organisms, allowing us to transfer genes between organisms, modify pre-existing genes, and even create entirely novel genes by linking up the appropriate nucleotides into any desired sequence.

Given that the genome contains all the information needed to make any organism into what it is, the potential of genetic engineering can hardly be overstressed. Ultimately it offers us almost complete control over the nature of life on the planet. Of course at the moment the things that can be done are still somewhat restricted – genes can only be transferred between certain organisms, the expression of engineered genes is sometimes difficult to control, the possible modifications and novelties are sometimes limited. But all the time the barriers between our present abilities and complete mastery are being steadily surmounted.

If viruses are accepted as "organisms" then they are clearly the simplest organisms of all, so we might reasonably expect that the viruses might be among the first organisms to be re-designed by genetic engineering. This is proving to be the case. For example, various researchers are trying to change vaccinia virus, of anti-smallpox fame, into a versatile live virus vaccine effective against other diseases. The basic approach is very simple – purify the genes that encode antigenic proteins of other viruses and then "stitch" them into the genome of vaccinia virus. This will produce a "recombinant" vaccinia virus whose genome will hopefully now produce proteins native to the target virus. The gene encoding the surface antigen of hepatitis B virus has been inserted into the vaccinia virus genome in this way; the hepatitis virus protein was produced in cells infected with the recombinant virus, and vaccination with the virus successfully protected chimpanzees against hepatitis B.[10]

Other scientists have put the genes for influenza virus or herpesvirus coat proteins into the vaccinia virus genome, again producing promising recombinant virus vaccines; and it might also be possible to put coat protein genes from *several different* viruses into *the one* recombinant vaccinia virus, allowing one virus to perhaps protect against several dangerous viral diseases.

The possibilities in vaccine design opened up by genetic engineering are truly unlimited. Obviously vaccinia virus is not the only candidate for a safe virus into which the coat protein genes of more dangerous viruses could be inserted. Over the next few years many viruses that themselves cause only trivial infections will be closely examined for their potential to be changed into recombinant virus vaccines. Other scientists are investigating completely different approaches such as removing the genes that make a virus dangerous and leaving the ones that allow it to stimulate an immune response; or creating hybrids of two dangerous viruses that retain the immune-stimulating properties of both and the dangerous properties of neither.[11] Overall, the previously haphazard and fortuitous approach to the production of attenuated live virus vaccines is going to be steadily replaced by the methodical alteration of viruses towards precisely defined aims.

As always, however, there are possible dangers and problems; particularly the fear that any viruses scientists "create" might be able to integrate into cellular DNA and cause cancer, or initiate other poorly understood diseases. The understandable excitement of scientists when presented with all the new possibilities will need to be restrained by rigorous checks to ensure safety.

Genetic engineering might well provide us with the "ideal" vaccines that previously existed only in vaccine designers' dreams, or alternatively its value might turn out to be rather more limited. But we can be sure of one thing – it *will* make an impact. Genetic engineering will permanently change the way in which many vaccines are made.

CHAPTER 12

Cures – the dawn of an era?

Whisky, honey, blackcurrant juice, herbal teas and vitamin pills – these are just a few of the "remedies" that people turn to when they fall victim to the common cold. Few other failures provoke more public criticism of modern medicine than its inability to help sufferers from that brief, trivial, but most irritating infection. Failure to conquer the viruses that give us colds is only the most obvious aspect of the inadequacy of modern medicine against viral infections in general. In an age when antibiotics can combat almost any bacterial infection, doctors are still frequently helpless once a virus has taken hold. It is often just a case of keeping the patient as comfortable as possible, perhaps using drugs to relieve such symptoms as pain and fever, and waiting for the body's own defences to fight the virus off. This helplessness has emphasised the need to develop and deploy vaccines against as many viral infections as possible, allowing people to be protected against diseases for which no cures are available. But the traditional family doctor's cry of "there's little we can do against a virus" may soon be completely out of date.

Already there are a few anti-viral drugs in use (although most are of limited efficacy and restricted range) and the signals coming from the world's anti-viral research laboratories have never looked more promising. A host of different chemicals targeted at various stages of the viral life-cycle are currently under trial. In 1982 Profesor Chien Liu of Kansas University felt sufficiently confident about such trials to declare "the dawn of an anti-viral era is approaching".[1] Hopefully the late 1980s will see the full light of that dawn, with victims of the viruses being increasingly offered not just symptomatic relief and sympathy, but also possible cures.

The failure to develop effective anti-viral drugs has traditionally been blamed on the intimate relationship between a virus and its host cell. Since viruses use much of their host cells' metabolic machinery to mediate viral multiplication, it has been difficult to find drugs that can interfere with viral multiplication while leaving healthy cells unharmed. This standard excuse, however, is looking increasingly flimsy as more is learnt about the molecular details of the viral life-cycle. Many viruses produce some of their own enzymes, for example, which might be suitable targets to attack with enzyme-specific drugs. The development of drugs that interfere with crucial viral enzymes while leaving host cell enzymes unaffected, might put us well on the way towards cures for the viral diseases concerned. This type of anti-enzyme attack is behind the success of "acyclovir" – an anti-herpesvirus drug produced by the Burroughs Wellcome Company, and one of the most promising anti-virals currently on the market. Taking a look at how acyclovir works should give you a good feel for the anti-enzyme approach in general.

Acyclovir

Acyclovir owes its anti-herpesvirus activity to strong similarities between its own chemical structure and that of the basic building-blocks of DNA (see figure 12.1). We saw in chapter 2 that DNA is made out of four nucleotides, each comprising a base, a sugar ring and a phosphate group. Acyclovir has the base "guanine" attached to a short chain of atoms that resembles the top of a sugar ring. This makes it similar not only to nucleotides, but also to "nucleosides" (nucleotides minus their phosphate groups) and some other chemical intermediates formed during the manufacture of DNA. Since it looks very similar to many of the chemicals involved in making DNA, acyclovir can sometimes take the place of these chemicals and interfere with viral DNA replication in the process.

Within herpesvirus-infected cells, acyclovir is converted into an active form by a specific herpesvirus enzyme that mistakes the drug for some DNA precursors. This active form then specifically interferes with the viral enzyme responsible for copying viral DNA (see box 12A and figure 12.1). The normal enzymes of the host cell can neither activate acyclovir, nor be adversely affected by the

BOX 12A – ACYCLOVIR

Acyclovir looks very like two nucleosides used to make DNA, known as "thymidine" and "deoxycytidine". Before they can be used to make DNA these nucleosides must first have a phosphate group added to them to convert them into nucleotides. Herpesviruses produce a specific viral enzyme (known as "thymidine/deoxycytidine kinase") that catalyses this phosphate addition. This enzyme can mistake acyclovir for a natural nucleoside and add a phosphate group to the acyclovir molecule. This addition of a phosphate group by the herpesvirus enzyme is the first step in the conversion of acyclovir into its active anti-viral form (see figure 12.1A). In the next step, enzymes native to the infected cell add two more phosphate groups to generate "acyclovir triphosphate". This again takes place because the relevant enzymes mistake acyclovir for the natural building-blocks of DNA.

Acyclovir triphosphate is the form of the drug that actually interferes with herpesvirus multiplication (see figure 12.1B). As you can see from the figure, acyclovir triphosphate looks very similar to "deoxyguanosine triphosphate", which is the form in which the nucleotide deoxyguanosine phosphate (G) must be before it can be linked up into DNA. This similarity allows acyclovir triphosphate to bind to the viral enzyme (known as "DNA polymerase") that links the four nucleotide building-blocks into new chains of DNA. By binding to the viral DNA polymerase enzyme acyclovir triphosphate can interfere with its activity and so prevent new copies of the viral DNA from being made.

Acyclovir triphosphate probably interferes with the DNA polymerase enzyme in at least two distinct ways. Firstly, simply by competing with deoxyguanosine triphosphate for a binding site on the enzyme it can prevent the proper incorporation of G nucleotides into DNA. Secondly, the drug might itself be mistakenly added onto the end of a growing DNA chain, in place of a G nucleotide. If this happens then the structure of acyclovir will prevent further growth of the chain from taking place, since acyclovir lacks the necessary chemical groups for the next nucleotide to latch on to. Some research also suggests that the interaction of acyclovir triphosphate with herpesvirus DNA polymerase might result in the *permanent* inactivation of the enzyme – a possible third and perhaps crucial aspect of acyclovir's activity.

Acyclovir specifically stops *viral* DNA replication because the acyclovir triphosphate cannot bind to and interfere with the

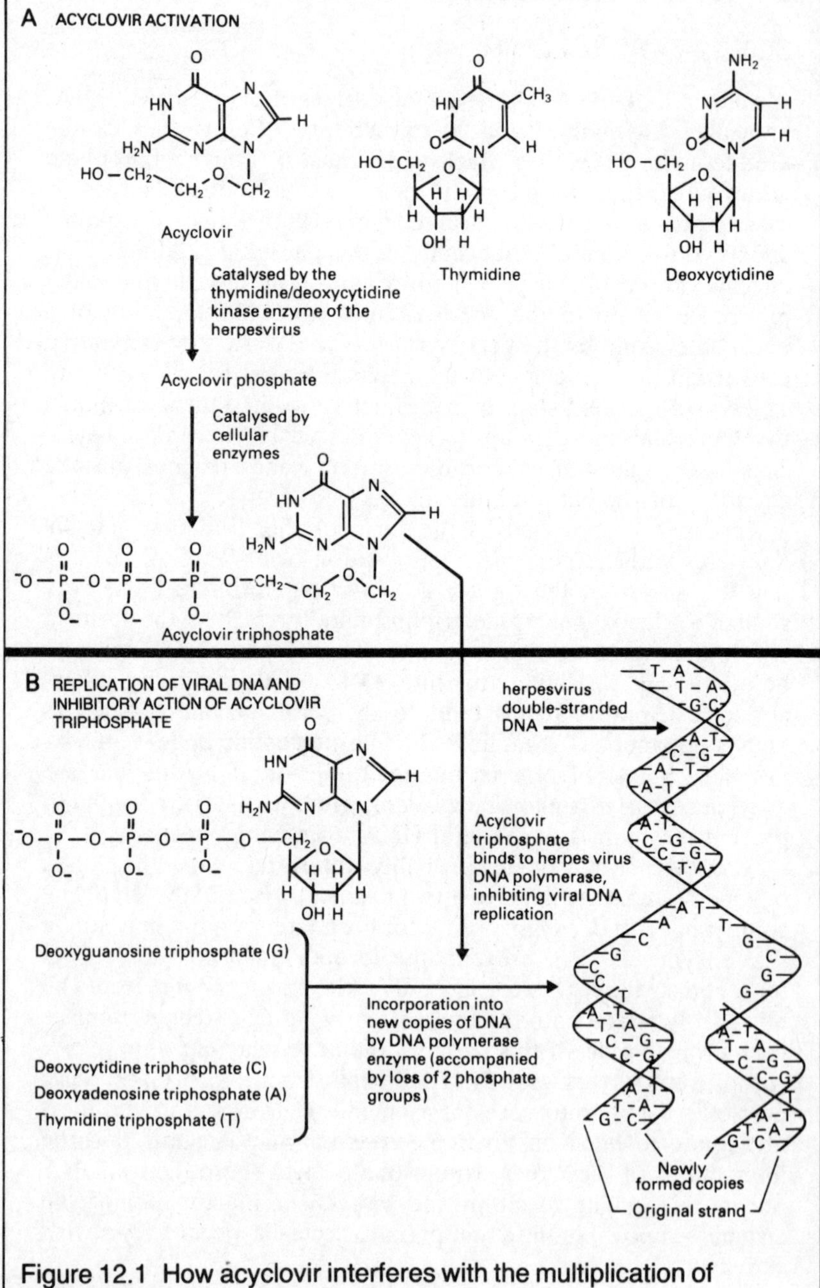

Figure 12.1 How acyclovir interferes with the multiplication of herpesviruses
(*Copyright © 1983 IPC Magazines Ltd*)

cellular DNA polymerase enzyme (which replicates cellular DNA) as effectively as it can interact with the viral enzyme. Also, the enzymes of healthy uninfected cells cannot efficiently perform the initial phosphate group addition that begins the conversion of acyclovir into its active form. So the drug is only converted into the active form within cells infected with a herpesvirus.

drug in its active form. So *its specific ability to interact with viral enzymes* makes acyclovir an effective and specific anti-viral drug.[2]

Acyclovir is administered as an ointment, in tablet form, or by injection directly into the bloodstream. It cannot be described as a "cure" for most herpes infections; but against diseases such as genital herpes, shingles, herpesvirus encephalitis and so on, it is proving to be of great benefit in accelerating healing, reducing the severity of symptoms and preventing recurrences.[3] Acyclovir is only one of a large number of similar chemicals being developed as possible anti-virals. All of these potential drugs share with acyclovir a similarity to the building-blocks of DNA and RNA (technically, they are all "nucleoside analogues"), and like acyclovir they work by being mistaken for these natural precursors of the nucleic acids. Of course in order to be specific anti-viral agents they must at some stage be dependent on viral enzymes to convert them into an active form (ensuring that they only go to work within virus-infected cells), or else they must selectively bind to and interfere with viral enzymes directly. The success of acyclovir suggests a bright future for this class of anti-virals. Laboratory trials have already shown that some of the newer nucleoside analogues under development might prove to be considerably more effective than acyclovir.[4,5]

Other avenues

Interfering with enzymes involved in viral multiplication is certainly not the only avenue of anti-viral attack. One appealing alternative is to stop viruses from becoming active within cells in the first place. Already there is a drug on the market, called "amantadine", that acts against influenza viruses at this stage.[6] Amantadine is a cage-like hydrocarbon molecule and exactly how it works is unclear. It probably stops newly entered viruses from uncoating and becoming active, but whatever its mechanism of action the important thing is that it *does* work. Its main value is in *protecting* people against infection during an epidemic, but it can also lessen the symptoms of influenza if given within 2 days of the symptoms first appearing.

Another anti-viral whose mechanism of action is currently a mystery is being developed by Roche for use against the common cold.[7] Presently known only by the codename "Ro–09–0415", this experimental drug has an interesting history that will please proponents of some of the commonly disregarded "folk" remedies for illness. In search of anti-cold agents, Roche turned their attention to a Chinese herb called *Agastache folium* – a traditional remedy for the cold. Roche were able to extract a chemical from the herb that seemed to be highly effective at stopping rhinoviruses (which cause many colds) from multiplying within cultured cells. They went on to manufacture a closely related derivative of the herbal compound which has similar anti-viral effects while being more readily absorbed into the bloodstream, and it is this derivative that has been labelled Ro–09–0415. The drug seems to bind to the surface of rhinoviruses and thus somehow prevent them from establishing an infection. Trials are presently only in their very early stages and no doubt many people will wish them well. A cure for many or even most bouts of the common cold would probably do more for the reputation of modern medicine than any number of advances against more serious but less common ailments.

For the present, though, hopes of any real "cures" for most viral infections remain to be realised. Amantadine and nucleoside analogues such as acyclovir are currently the only drugs routinely available to combat the viruses; and however promising their successes have been they still remain rather limited. We will shortly be considering some of the experimental approaches aimed at improving the situation, but in the meantime we should remember that the body itself usually mounts a highly effective anti-viral defence. Another approach to curing viral disease is to try to exploit the ready-made remedies contained within the body. Two particular approaches merit our attention – the administration of antibodies to boost a patient's own immune response, and the anti-viral potential of interferon.

Antibodies and interferon

The protective effect of administering purified antibodies to someone suffering from a viral infection is easy to understand. Provided the antibodies bind specifically to the virus involved they can either precede or augment the patient's own antibody response. Purified antibodies have been used against serious viral infections for some years, but one of the main problems with this form of treatment has been the difficulty in obtaining large quantities of suitable antibodies. At first animals such as horses were used as a source of supply, but the non-human antibodies obtained could often induce harmful allergic reactions. More recently, the required antibody supplies have been obtained from donated human blood, allowing effective treatment of such infections as rabies, tetanus and hepatitis B. The new developments in biotechnology discussed in the previous chapter might make antibody production much easier in the future, but the practice of administering antibodies is a far from perfect approach to curing viral disease. For one thing, its effectiveness is often restricted to either before or immediately after an infection has properly taken hold, making it considerably less useful than drugs that could attack an infection at any stage. Secondly, specific antibody preparations are needed against each particular virus, which again makes this approach less useful than generally active drugs. The administration of antibodies might well be an important part of the hospital treatment of serious viral diseases in the future, but therapies available on prescription for people to take away with them will need to come from other avenues of attack. One of the most talked about possibilities is interferon – the natural protein that has stolen most of the headlines recently from drugs such as nucleoside analogues which might well turn out to be more useful.

Interferon (or more strictly the interferon*s*) has the great advantage of being effective against a wide range of viruses. As we saw in chapter 7, it can convert cells into an anti-viral state in addition to its still mysterious ability to modulate various aspects of the immune response overall. Much of the media interest in interferon has centred on its possible use as an anti-cancer drug, but its anti-viral activities might prove to be of greater benefit. The potential of interferon as a treatment for human viral disease was demonstrated as early as 1962, when it was found to lessen the localised skin disfiguration produced by smallpox vaccination.[8] But testing it against serious illness in a rigorous way was for many years severely restricted by lack of supplies. Of course the genetic engineering revolution has recently changed all that, with large quantities of pure and relatively cheap interferon being produced by microbes into which the interferon gene has been artificially inserted.

Many trials of the effectiveness of interferon against many different viral diseases are currently under way. The final verdict on interferon's efficacy as a general anti-viral is not likely to be in for several years yet, but some of the early results have prompted both hope and caution.[9] Interferon has "reduced" the spread of herpesviruses and the associated symptoms in some patients suffering from shingles. It has proved of "some benefit" in the treatment of chronic hepatitis B. It has "reduced" the frequency and severity of attacks of genital herpes, and so the results go on. You will gather from the quotations that there is certainly no justification at the moment for hailing interferon as an anti-viral panacea.

One of interferon's most publicised early "victories" was over the common cold. The British Medical Research Council's Common Cold Research Unit, for example, has found that interferon taken as an intranasal spray can provide highly effective protection against the cold. This discovery obviously aroused great interest and even excitement, but in 1983 the same research team was reported to have abandoned its trials because of the side-effects associated with the use of interferon. The interferon apparently caused nose bleeds and discomfort, and it has also been associated in other trials with such diverse side-effects as fever, headache, hair loss, growth retardation and suppression of the developent of blood cells.[10] This protein hailed as a possible cure for cancer has even been suspected of *causing* cancer in some cases.[11] So the early hopes that interferon, as a *natural* anti-viral, would be free of the harmful side-effects commonly found with man-made drugs, are proving to be unfounded.

Some research groups, however, feel that the side-effects are generally insignificant or will eventually be overcome. The interferon story has a long time to run before any final conclusions can be drawn – dosages will need to be altered, routes of administration varied and alternative combinations of all the different interferons tried out. Genetic engineering also holds out the possibility of constructing modified interferons, unlike any found in nature, which might retain the wanted anti-viral effects without producing any unwanted side-effects. It will be fascinating to see how the long-running saga of interferon finally turns out. The fate of several young biotechnology companies (and also, of course, thousands of patients) might well be decided by the outcome.

Even if interferon itself does not live up to initial expectations, however, the increasing knowledge about how it works might lead on to the development of useful drugs. We saw in chapter 7, for example, that viral double-stranded RNA seems to induce the manufacture of interferon within cells. So might it be possible to treat viral diseases with such *inducers* of interferon rather than with interferon itself? This idea has already been tried out in clinical trials of a simple synthetic RNA known as "poly I:C". Interferon production is certainly switched on by this compound, but its anti-viral effects overall have been rather disappointing.

But interferon inducers are not the only possibilities thrown up by research into interferon's action. In chapter 7 I said that interferon stimulates a cell to manufacture various anti-viral proteins, without going into any details. In fact, it appears that one of the first effects of interferon is to increase the production of a small molecule known as "2,5-A" (actually 2′,5′ -oligoadenylate). The 2,5-A then itself switches on an enzyme responsible for destroying viral messenger RNA. Synthetic analogues of 2,5-A have been tested to see if they might be effective anti-virals and some of the early results have been promising. Various analogues have been shown to interfere with viral multiplication in cultured cells; and while trials in cultured cells fall a long way short of demonstrating that these drugs will be safe and effective in the treatment of human disease, they certainly suggest that the potential of 2,5-A analogues is worth pursuing.[12]

Finally, before leaving the anti-viral possibilities offered by the body's own immune response, I should mention a new drug called "imunovir". Imunovir is a combination of three different and rather simple chemicals that seem to stimulate the immune system into action, allowing it to fight off infection more successfully. So its main effect is not to attack viruses directly, but simply to give a boost to the immune response (which is often depressed during the course of a viral infection). How imunovir works is not really known, but it has been shown to be of some benefit in the treatment of herpesvirus infections, and investigations into its effect on other important diseases (including the infamous "AIDS") are currently under way.[13] The UK manufacturers of imunovir (Edwin Burgess Ltd.) believe that it "marks the beginning of a new era in the treatment of viral disease", but it will be some time until we know whether or not such confidence is justified.

Diverse possibilities

Nucleoside analogues and interferon form the mainstream of current investigations into anti-viral therapies, but many other possibilities are being investigated.[14] Roche's anticold drug, interferon inducers, 2,5-A analogues and immunostimulators such as imunovir are just four of many. Let's take a brief look at a few of the others.

Professor Purnell Choppin of the Rockefeller University in New York is busy making peptides that look like the crucial regions of viral coat proteins involved in binding to the receptor proteins carried by a host cell. His hope is that these peptides might compete with the viral proteins for the binding sites on the receptors, thus preventing or at least greatly reducing the entry of viruses into susceptible cells (see figure 12.2). He has found that when appropriate peptides are added to cultured animal cells then the ability of some viruses to infect the cells is considerably reduced.[15] The crucial animal trials required to show whether or not the same protection is afforded when the peptides are administered to a living host are currently under way. An alternative possibility, of course, might be to manufacture peptides that mimic the binding sites of the cell receptors and thus bind to invading viruses to prevent their attachment to these receptors (see figure 12.2).

Paul Zamecnik and Mary Stephenson of Harvard University have linked up 13 nucleotides to produce a synthetic "oligonucleotide" (oligo = a few) that is complementary to a short region found at both ends of a cancer-causing retrovirus's genome. Their hope was that the oligonucleotide would bind by base-pairing to the viral genetic material, hopefully preventing the viral genes from working properly (see figure 12.3). When they added their oligonucleotide to cells infected with the retrovirus the production

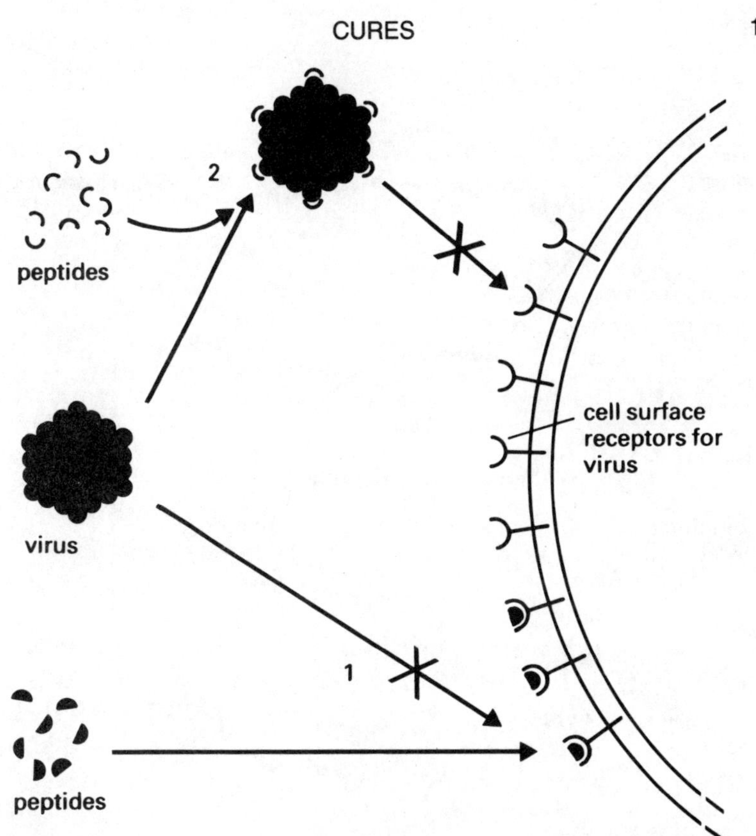

Figure 12.2 How peptides might be used as anti-viral drugs. Synthetic peptides designed to mimic crucial regions of a virus's coat proteins might bind to the cell surface receptors and so block the binding of the virus 1). Alternatively, peptides designed to mimic parts of the receptors might bind to the surface of a virus and so prevent the virus from binding to the real receptors 2)

of new virus particles was strongly inhibited.[16] As always, we should not get over-excited about such results until they have been repeated in live animals, and other research groups have failed to demonstrate any anti-viral effects with various other synthetic oligonucleotides.

Before viruses are finally assembled prior to leaving an infected cell, some viral proteins have carbohydrate groups attached to them by enzymes native to the infected cell. These "glycosylation" enzymes thus convert viral proteins into glycoproteins, and various chemicals known as "glycosylation inhibitors" have been found to

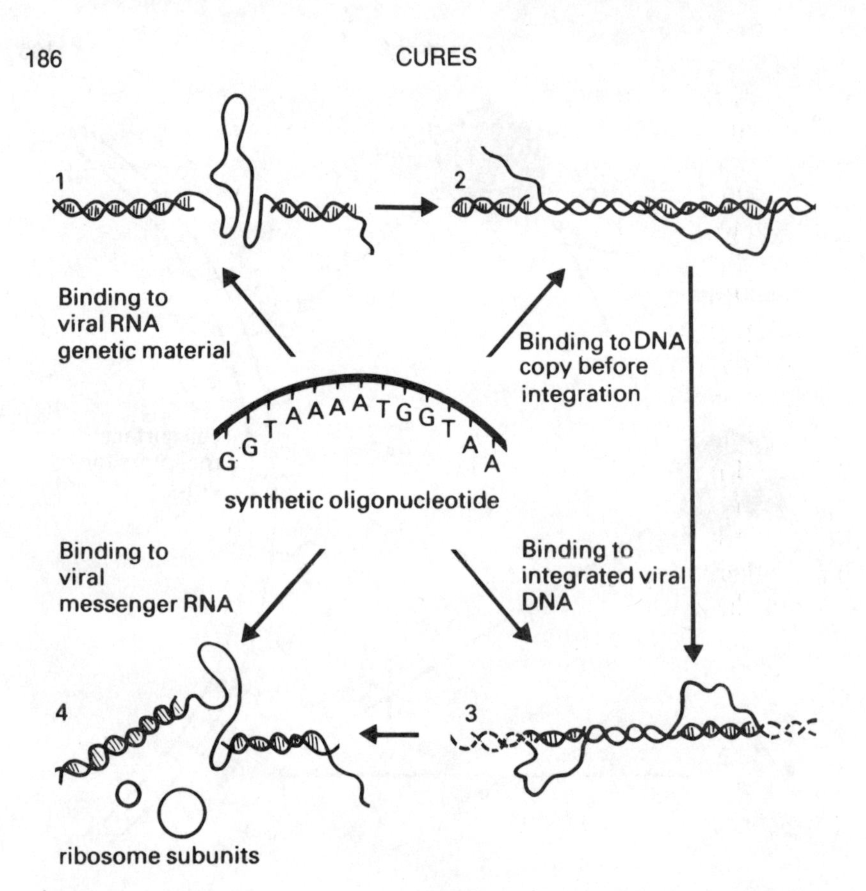

Figure 12.3 A synthetic oligonucleotide that is complementary to end regions of a retrovirus's genetic material could inhibit the life-cycle of the virus in four ways: Binding to the newly entered viral RNA might prevent copying into DNA 1); Binding to the DNA copy might prevent integration into host cell DNA 2); Binding to integrated viral DNA might prevent copying into messenger RNA 3); Binding to messenger RNA might prevent ribosomes from binding and so prevent protein synthesis 4). Other synthetic oligonucleotides might be able to bind to and interfere with the genetic material of viruses other than retroviruses *(Copyright © 1983 IPC Magazines Ltd)*.

interfere with this process. "2-deoxyglucose" and "glucosamine" are two particularly promising glycosylation inhibitors that might eventually be used in effective anti-viral preparations, but how they work is not really known.

Other potential anti-virals that work in still mysterious ways are ionic complexes of either lithium or zinc. A lithium-containing ointment, for example, can reduce the pain of genital herpes and cut down the number of new viruses released from infected tissue,[17] while lozenges containing a compound of zinc can apparently shorten the average duration of the common cold.[18]

So the world of anti-viral research is currently alive with a welter of possibilities, some of which will hopefully crystallise into highly effective therapies. Ever since the discovery of penicillin began the conquest of bacteria, a similar anti-viral "penicillin" has become the drug researcher's holy grail. But all the available evidence suggests that there will not be one dramatic penicillin-like breakthrough in the development of effective anti-virals. Instead, there will probably be a gradual wearing down of the destructive effects of various viral diseases, with a wide variety of drugs becoming increasingly effective against specific types of virus. At the moment several different compounds are of "some benefit", but we have no really reliable cures. Considerable problems remain between the present state of affairs and the cures that everybody wants; but let's hope that Professor Liu has got it right, that the problems will be solved and that the long-awaited anti-viral era really is on its way.

CHAPTER 13

Mysteries – many puzzles remain

Every month the science libraries of the world receive the latest influx of knowledge about the often weird and wonderful activities of the viruses. Countless journals arrayed on multi-layered and ever-lengthening display racks contain more information about viruses than any one person could ever have time to read. Some of these journals are devoted entirely to studies of the viruses, others include virology as part of their coverage of microbiology as a whole, but even within general science journals the viruses manage to claim a large share of the available space. All this interest is well warranted – in addition to the obvious medical incentives to study virology, the viruses are also widely used as "simple" model systems pointing us towards greater insight into the activities of their much more complex host cells. The preceding chapters have provided just a very general overview of the impressive mass of information about the viruses gleaned from over 90 years of research. The activities of some individual viruses have been studied in such detail that entire books have been devoted to them. One such book, called *The Bacteriophage Lambda*, is entirely concerned with the small bacteriophage we met briefly in chapter 5.[1] But despite the great edifice of facts that *are* known about the viruses, there remain an awful lot of things that aren't known. In this chapter I want to tell you about just a few of the specific mysteries currently taxing the brains of the world's best virologists.

Some of the most relevant mysteries concern the diseases which viruses can cause. In the early years of virology the familiar and often dramatic viral illnesses such as smallpox, measles, mumps, chickenpox and so on were fairly easily identified and readily attributed to the viruses. Such diseases are caused by acute viral infections which quickly destroy the infected cells. But as more and more of the traditional virus diseases are being fought off with vaccines, interest is turning to more mysterious illnesses which might also be caused wholly or partly by viral infection. For example, viruses have been accused of causing such serious illnesses as diabetes, rheumatoid arthritis, atherosclerosis, multiple sclerosis and various other brain diseases including schizophrenia; but in some cases the evidence is very flimsy and almost always factors other than viral infection are also involved.

But these are all "old" diseases, in the sense that they have occurred, and we have known about them, for a very long time. There must also arise, from time to time, completely new or at least drastically altered viral diseases that have not been experienced before. Our relationship with the viruses is after all a dynamic, ever-changing one – with viruses altering in response to the challenge of modern medicine, the new opportunities presented by our changing lifestyles, or perhaps spontaneously acquiring the ability to infect human cells in addition to the animal cells they might previously have been restricted to. The past few years have seen the alarming rise of a "new" disease called "acquired immune deficiency syndrome", usually known simply as "AIDS". The acronym "AIDS" has become familiar to everyone, thanks to headline-writers the world over, and AIDS is rapidly displacing genital herpes from its position as the world's number one sexual "scare" disease. Its prevalence among homosexual communities has led some people to attribute it to the avenging wrath of God, while its spread into the heterosexual population has raised fears that it might become a widespread new "plague".

Amongst all the concern and speculation about AIDS, virologists (while of course sharing in the general concern) can see in the rise of the disease an opportunity. If AIDS is caused by a virus (and all the evidence suggests that it is) then it may be possible to follow in detail the progress of a novel viral disease from its earliest outbreak to its final conclusion. Hopefully the attentions of virologists will ensure that that conclusion is as happy as possible from a human point of view, with the virus being either quickly conquered or at least effectively controlled; but at the moment such a happy conclusion is certainly not assured.

AIDS[2]

In the late 1970s some young and highly promiscuous male homosexuals in America, particularly in San Francisco and New York, began to fall ill in a very mysterious way. They began to suffer and eventually die from infections that are normally harmless or usually cause only mild disease, and also from some bizarre types of cancer rarely seen in the population at large. The two most common ailments were pneumonia caused by a specific bacterium (*Pneumocystis carinii*) and a skin cancer known as "Kaposi's sarcoma", but these were by no means the only problems. A host of other viral and bacterial infections began to develop in the victims and progressively get worse; while in place of Kaposi's sarcoma there sometimes appeared other cancers which (like Kaposi's sarcoma) were strongly suspected of being at least partially viral in origin. All of these early signs suggested that the patients were falling victim to what are known as "opportunistic" infections – normally mild infections that become much more serious when the victim's immune system is impaired in any way. So the young homosexuals appeared to be suffering first of all from a deficiency of the immune system, which was in turn providing the various infections with the "opportunity" to spread unhindered and eventually cause death.

The suggestion that immune-deficiency was at the root of the problem was confirmed when blood from the early victims was examined in the laboratory. The proportions of some of the cells responsible for immunity were all wrong. Particularly prominent was a decline in a subset of T-cells known as "helper" cells, which help the rest of the immune response cells to mount an effective challenge against an infection. Interferon production was also abnormal, the activity of natural killer cells was reduced and there seemed to be less immune system cells present overall than there should be in a healthy individual. So in general the immune system of these unfortunate homosexuals was clearly defective, they were suffering from an "immune deficiency syndrome", or in other words they had AIDS.

Very soon after the first cases of AIDS were discovered it became apparent to doctors interested in the condition that they might have a major problem on their hands. What was most alarming, apart from the regularity with which AIDS caused death, was the exceptionally fast increase in the rate at which new cases were appearing. The very first American case probably arose in 1978. Since then the number of cases appears to have doubled every 6–8 months, producing an early 1985 figure of well over 8000 cases in America and over 1000 in Europe. A few moments with pencil and paper should convince anybody that the *rate* at which AIDS is spreading is truly alarming. Starting from an early 1985 figure of 9000 and doubling at the rate of once every 9 months (a currently conservative estimate), would produce 100,000 cases sometime in 1987, and over 10 million sometime in the early 1990s. Of course eventually the rate at which a new disease spreads can be expected to stabilise, and stability might be reached at a level far below the figures calculated above. The rate of increase may well be levelling off in America already, but it will be some time before any realistic limits can be put on the amount of damage that AIDS might eventually do.

A comfort to many people in the early days of the AIDS epidemic was the hope that the disease might be restricted to homosexuals. Such complacency was unjustified, for although AIDS appears to have arisen within the homosexual community (certainly in the West – see later), it then began to spread to other groups. Cases of AIDS have occurred in drug addicts who take their poison by the intravenous route, female partners of AIDS sufferers and recipients of transfusions of blood or purified blood products which have been contaminated by AIDS-infected donors. All the evidence suggests that we are all potential victims of AIDS, it can be transmitted during both heterosexual and homosexual intimacy as well as within contaminated blood, and indeed the agent that causes AIDS seems to be passed on in much the same way as the virus that causes hepatitis B. The prospect of AIDS ever becoming as widespread as hepatitis B infection (with an estimated 150 million carriers worldwide) is truly appalling.

Having destroyed the complacency of anyone tempted to dismiss AIDS as "the gay plague", I should tell you a bit more about the prospects for its victims. Put simply, the outlook is very grim indeed. The overall death rate so far is about 50 per cent, but this does not mean that 50 per cent of AIDS victims survive – simply that many of the most recent victims have not yet had time to die. The figures look much worse when you realise that 70 per cent of all victims have died within 1 year of contracting the disease, so most people with AIDS who are alive as I write will be dead by the time this comes to be published and read. Perhaps the single most alarming statistic about AIDS is that no victims have ever recovered their normal powers of immunity, either spontaneously or after attempted treatment. There have been no remissions and so far there is no cure. All that doctors have been able to do so far is to treat the infections to which an AIDS sufferer succumbs, perhaps fighting off one or even a few; but it seems that in almost all cases one infection (or one of the cancers) eventually proves unconquerable and the victim then dies.

Clearly the threat of AIDS justified an intensive research effort to try to identify its cause and develop either cures or means of control. The state of play in that effort can be briefly summarised as follows – AIDS is certainly infectious, and is almost certainly caused by a virus which was only properly identified in 1984. The identification of the culprit virus will allow blood to bc screened for contamination, should allow earlier and more reliable diagnosis, and will hopefully lead on to the production of successful vaccines to protect against the disease. The attempt to find effective treatment for use once the diagnosis has been confirmed has had no dramatic successes, although several approaches have shown some promise.

The simplest evidence suggesting that AIDS is caused by some sort of infectious agent is the classically epidemic-like spread of the disease through a close-knit population – the homosexuals of San Francisco and New York. Further evidence comes from the ability of AIDS to be transmitted via contaminated blood. For example, hundreds of haemophiliacs have contracted AIDS from purified blood-clotting proteins produced from donated blood; and many people have caught the disease during blood transfusion. The transmission of AIDS between drug addicts also supports the involvement of some blood-borne infectious agent.

The similarities between AIDS transmission and that of hepatitis B, mentioned earlier, are certainly very close. Hepatitis B, remember, can spread not only via contaminated blood but also along with semen and saliva during intimate sexual contact. The semen and saliva pathways may well be the ones used by the AIDS agent during sexual contact, which would explain the initial outbreak of the AIDS epidemic within a highly promiscuous population. It is quite possible that the homosexuality of most AIDS victims is an incidental factor, secondary to their crucially important promiscuity.

The links between AIDS and hepatitis B led to the suggestions first of all that AIDS might also be caused by a virus (viruses can certainly suppress the immune system), and secondly that the hepatitis B virus might itself be responsible. Many AIDS victims do suffer from hepatitis B, but not all; and in any case it is obviously difficult to distinguish between infections that might *cause* AIDS and those that develop as a *result* of the immune deficiency found in AIDS victims.

A great many viral infections have actually been linked to AIDS and considered as potential causes, but for a long time none of them emerged as really convincing candidates. The best candidate throughout 1983 was the retrovirus known as HTLV: known to cause human leukaemia, known to infect the T-cells that seem most affected by AIDS, and linked to many cases of AIDS. But there were problems with blaming the known types of HTLV for AIDS. For one thing these viruses did not *kill* the T-cells they infected, but instead could convert them into immortal cancer cells. This did not seem to fit in with the drop in the number of viable "helper" T-cells found in AIDS.

Confusion and mystery remained the hallmarks of the AIDS story until early in 1984, when an American research team led by Robert Gallo of the US National Cancer Institute, and a French research team led by Luc Montagnier of the Pasteur Institute in Paris, reported a crucial discovery. They identified a new virus which seems to be a really convincing candidate for the virus that causes AIDS.[3] The Americans named their putative AIDS virus as "HTLV-III", since they believe it belongs to the HTLV class of viruses. The French workers call *their* virus "LAV" (for "lymphadenopathy AIDS virus"), but both appear to be very closely related variants of the same basic AIDS-causing virus. It can be isolated from at least 30 per cent of AIDS victims, and from nearly 90 per cent of people suspected of having an early form of the disease. Antibodies directed against the HTLV-III/LAV virus are present in over 90 per cent of AIDS victims, and possibly all of them. And most importantly, when the virus is added to cultured human T-cells it tends to kill them rather than being able to make them cancerous and therefore immortal. So it is easy to understand how this virus may cause AIDS by infecting T-cells, killing them, and so producing immune deficiency.

The discovery of HTLV-III/LAV and its links to AIDS was hailed by much of the media with such clear-cut headlines as

"AIDS virus found". Such confidence was perhaps a little premature, for as I write absolutely conclusive proof is still lacking. Of course AIDS researchers face the same problem as scientists studying the links between viruses and cancer – how do you prove that a specific virus causes a particular serious human disease? The best answer would be to develop a vaccine against the virus and demonstrate that it can prevent the disease. The possibility of an anti-AIDS vaccine is just one of the benefits we can expect from the discovery of HTLV-III/LAV – assuming it really is the AIDS virus. Other likely benefits are the ability to screen donated blood for the virus to prevent the spread of AIDS to transfusion recipients or users of purified blood products; and efficient diagnostic tests that can detect the virus and so identify people at risk of developing AIDS at a much earlier stage than is now possible.

AIDS research is such a fast-moving field that by the time this book comes to be read reliable screening, early diagnosis and perhaps even successful vaccination might already be routine. One further area in which important advances might soon be made is the search for ways to help AIDS victims once they have contracted the disease. The drug we looked at briefly in the last chapter, called imunovir, might reasonably be expected to help since it works by stimulating the immune system. Might imunovir be able to boost an AIDS-afflicted immune system back up to its normal vigour? Early trials are proving promising. It has not yet been shown to help patients with full-blown AIDS, but imunovir has apparently been able to repair the damaged immune system of people suffering from what is believed to be an early precursor stage of the disease.[4]

Other promising signs have come from studies of a natural protein known as "interleukin-2", which switches on several aspects of the immune system known to be deficient in AIDS. Human interleukin-2 has been manufactured within bacteria using genetic engineering techniques, administered to blood cells isolated from AIDS victims and shown to increase the ability of these cells to mount an immune response.[5] Such "*in vitro*" studies fall a long way short of proving that interleukin-2 will help real live AIDS victims. Early trials on AIDS patients have produced no dramatic successes but further and more intensive trials are currently under way. Of course the more famous natural protein, interferon, is also known to influence the immune system and it too is being tried out on AIDS patients. But so far AIDS remains an incurable disease.

Whatever future developments there might be in the treatment and prevention of AIDS, the intriguing problem will remain of where has it come from in the first place? At present there are no firm answers to this question but one particularly interesting clue.

It has turned out that many cases can be linked in some way to two particular Third World countries – Haiti (linked to American AIDS cases) and Zaire (linked to European cases). The chain of AIDS infection can sometimes be traced back to people having recent contact with these countries, and the inhabitants of Haiti and Zaire also suffer from AIDS (or something very like it), where it attacks both men and women and is passed on during heterosexual contact. It has therefore been suggested that these (and perhaps other) areas of the Third World might be the "cradle" of AIDS, providing the base from which it has spread to the West. It might have arisen in the Third World quite recently, or else have existed there (possibly in a milder form) for quite some time. It certainly seems possible that an immune deficiency disease such as AIDS might have gone unnoticed in the Third World for some time – the people are generally less healthy, infection and malnutrition (both of which can themselves cause immune deficiency) are rife, and the surveillance of disease is much less efficient than in the West.

Could the rise of international travel have brought a fairly common disease in some parts of the Third World to the pampered population of the West, in whom it might perhaps be much more noticeable? Might the effects of the disease on its new victims be much more severe due to their lack of any "inbuilt resistance"? Has the virus responsible recently mutated into a much more dangerous form? Is it perhaps genuinely "new" to humans, having previously been restricted to other animals? These are just some of the many questions that remain to be answered by scientists trying to unravel the mysteries of AIDS.

Subtle and slow

Many of the other current mysteries in virology concern well-known diseases which might be at least partly due to viral infection, and there again might not. As already mentioned, the viruses may be involved in diabetes, rheumatoid arthritis, atherosclerosis, multiple sclerosis and several other disorders of the brain; and the involvement of the viruses (if indeed they are involved) is often likely to be subtle and/or slow. Scientists examining the possible role of viruses in these diseases often have to leave the relatively simple world of acute dramatic infections, and delve into the world of quiet or persistent infections and the much more subtle interactions between some viruses and the cells they infect.

Some of the most persuasive evidence incriminating the viruses in any of the diseases listed above concerns "type 1" diabetes, in which the pancreas fails to make enough of the hormone insulin. Insulin is normally manufactured by "beta-cells" found in the pancreas, from where it passes out into the bloodstream and (amongst other things) helps cells to take up glucose brought to them by the blood. If the beta-cells stop making insulin for any reason, the blood glucose levels will rise and the now diabetic patient may eventually fall into a coma and die. Fortunately, of course, the disease can be quite effectively controlled by the regular administration of pure insulin.

Several different lines of evidence suggest that viruses might have a role to play in the onset of type 1 diabetes, but one of the most direct involves the investigations of Dr. Ji-Won Yoon of the US National Institutes of Health. Yoon and his co-workers discovered a type of virus known as "coxsackievirus" in the pancreas of a child who had recently died of diabetes. They purified the virus and injected it into mice, whereupon the mice promptly developed diabetes due to the destruction of the beta-cells in their pancreas.[6] This result is commonly presented as "conclusive" evidence that the coxsackievirus caused the child's diabetes, and of course it suggests that viruses might often cause diabetes.

Many other types of viral infection (in addition to those due to coxsackievirus) have been linked to the onset of diabetes, but which viruses really *can* cause the disease and exactly how they might do it remains unknown. Obviously viral infection might destroy the insulin-producing beta-cells, but it might also simply shut down their manufacture of insulin while leaving the cells themselves intact. Even if the beta-cells are destroyed, the question remains of *how* they are destroyed. Is it the effect of infection directly, is it the result of the immune attack against infected cells, or can a viral infection induce an auto-immune response against the host antigens carried by beta-cells? Various studies have certainly indicated that auto-immune damage may well be involved.

Probably the most important point to emphasise is that viral infection does not seem to be the *only* possible cause of type 1 diabetes, indeed on its own it is probably not sufficient cause. Other factors such as chemical toxins acting on the beta-cells and the genetic make-up of the individuals concerned may also have a crucial role to play. So we arrive back at the sort of highly qualified links between viruses and disease that we have already met when considering the problem of cancer. Viruses apparently can cause diabetes, but they are not the only cause and the infection must coincide with other appropriate factors. So in a similar manner to the cancer viruses the "diabetes viruses" are probably best considered as just one of the various environmental insults that can participate in the cause of the disease, and probably only in some cases. But you will appreciate from all the "mights" and "probablies" that a lot of work remains to be done before the role of viruses in giving us diabetes becomes clear.

Two other very common diseases that still need a lot of work to be done on them before what causes them becomes absolutely clear are rheumatoid arthritis and atherosclerosis. In both cases links with viral infection have been made, but the evidence is often flimsy and certainly never clear-cut. Consider rheumatoid arthritis first of all. This is actually a disease which affects many parts of the body in a variety of ways, but its most obvious symptoms are painful joint inflammation (i.e. "arthritis") and the wasting away of muscles. Nobody knows what causes rheumatoid arthritis but much of the damage can apparently be blamed on the immune system. Of course auto-immunity might be involved, but equally the damage might be the result of a long fight between the immune system and some sort of persistent infection. In either case a viral infection could easily lie at the root of the problem.

The trouble with the infection theory of rheumatoid arthritis is that no viruses, bacteria or other micro-organisms can be *consistently* detected in the affected joints and tissues. Animals can certainly be given a form of arthritis by injecting micro-organisms into their joints, and some natural viral infections do produce brief bouts of joint inflammation. But such observations fall a long way short of proving that infection causes the long-term chronic damage found in human rheumatoid arthritis.

If an infection really is to blame, then viruses are the candidates favoured by many people, thanks to their ability to lie dormant in cells in forms that are often difficult to detect. But the disease may well turn out to be caused by bacteria, fungi, or not by infection at all.

As far as specific candidate viruses are concerned, some parvoviruses,[7] and the herpesvirus that causes glandular fever (Epstein–Barr virus),[8] have recently attracted the most attention. It has been claimed for example, that people with rheumatoid arthritis produce an antibody that binds specifically to an antigen found in Epstein–Barr virus-infected cells. But as a recent editorial in the medical journal *The Lancet* put it – the viral theory of rheumatoid arthritis "has been sustained more by faith and . . . analogy, than by tangible evidence".[9] The next few years might see this situation change dramatically, with verdicts against the viruses of guilty, partly guilty, or not guilty, all looking equally likely at the moment.

The possible connections between viruses and atherosclerosis are currently even more tenuous than their links to rheumatoid arthritis. Atherosclerosis is the medical name given to the hardening and thickening of the arteries that so often leads on to heart disease and strokes. It is actually a rather complex and still poorly understood phenomenon involving both the deposition of fatty material on the insides of arteries and the multiplication of some cells of the artery wall. An inappropriate diet is possibly the most celebrated factor involved in giving us atherosclerosis, with animal fats and cholesterol as the favourite culprits. But many other influences are probably important, perhaps including viral infection.

Professor Earl Benditt's research team at the University of Washington in Seattle have recently suggested that herpes simplex virus infection of the artery wall might be involved in atherosclerosis.[10] They base their case firstly on animal studies which suggest that a related herpesvirus infection in chickens can produce arterial damage similar to human atherosclerosis; and secondly on some very preliminary studies of diseased human arteries. When they looked at a sample of diseased arteries they found the genetic material of herpes simplex virus within the affected cells, but no genetic material belonging to other types of herpesvirus. Is it possible, they wonder, that herpes simplex virus (and possibly other viruses too) might be another factor to add to the list of risk factors that predispose us to atherosclerosis? In favour of this idea they point out that herpes simplex virus is known to persist within some infected cells and that it can induce the multiplication of certain cultured cells. So a persistent herpes simplex virus infection might be able to initiate and perpetuate the proliferation of artery wall cells found in atherosclerosis. At the moment the evidence barely makes a worthwhile case, but it presents yet another possible mystery about the viruses which will need to be resolved.

To finish this quick dash through some of the current medical mysteries in which *conventional* viruses may be involved, we must turn to various diseases affecting the brain. A great many mysterious diseases or "syndromes" occur within the brain, and the viruses may be involved in quite a few of them. There are actually a whole range of brain and nervous system disorders which have been collectively termed "post-viral neurological syndromes".[11] As the name implies, these conditions appear after various types of viral infection, but exactly how they develop and to what extent the viruses are involved remains a mystery. The diseases are described individually by such esoteric titles as "acute disseminated encephalomyelitis", "epidemic myalgic encephalomyelitis", "Guillain–Barré syndrome", "Refsum's disease", and so on. In many cases the affected nervous system appears to be the victim of the type of immune system-mediated damage considered in chapter 8, with the viruses somehow acting as the triggers needed to set the process off. How they might manage to do this remains to be fully determined.

One much-better-known brain disease that may well be caused by the viruses is multiple sclerosis, in which the fatty myelin sheath surrounding nerve cells is progressively destroyed. This damage to nerve cells interferes with their transmission of nerve impulses, causing the victims of multiple sclerosis to suffer from a variety of progressively worsening symptoms such as unsteadiness, poor co-ordination and double vision. We have already seen that measles virus can break down myelin when it persists to cause SSPE. Viruses are also blamed for the demyelination seen in multiple sclerosis, although much of the evidence is still confusing and inconclusive. It will be most appropriate here for me just to summarise the current situation. Good evidence suggests that multiple sclerosis is at least triggered by some sort of infection, and the viruses are the favoured candidates. No particular virus can presently be held responsible, and it is possible that several different viruses may be involved. It is suspected that within genetically susceptible individuals, viral infection(s) sets in motion a slow progressive attack by the immune system on myelin, with the virus either persisting as a permanent trigger of immunity or disappearing after initiating an auto-immune response that remains effective for life.[12] Almost all the details remain to be confirmed and properly characterised.

Prions, virinos, viroids

Mysteries abound, then, concerning the involvement of the viruses in many important and poorly understood human diseases; but so far, when considering mysterious illnesses that behave in some way as if they are caused by viral infection, I have looked at only the *conventional* viruses as possible causes. Might some *unconventional* virus-like infectious agents cause some of the various unexplained human diseases? This possibility is certainly being taken seriously, with several bizarre new types of infectious "organisms" being suggested, particularly in an effort to explain some slow degenerative diseases of the brain.

"Jakob–Creutzfeld disease", for example, is a rare human illness which behaves in many ways like an infectious viral disease. It gains a foothold on the brain and then slowly causes the degeneration of nervous tissue, presenile dementia and eventually death. But one startling fact suggests that a conventional virus cannot be responsible – the immune system is apparently not mobilised at all to fight off the infection. The diseased tissue does not become inflamed, no antibodies directed against an appropriate infectious agent appear to be formed, and so on. There is a well-studied parallel of this disease that occurs in sheep and is known as "scrapie". Scrapie, and the scrapie agent, have been used as the model system representing similar human infectious diseases such as Jakob–Creutzfeld disease, but progress in understanding all the strange scrapie-like diseases has been severely restricted by the failure either to properly purify the agent responsible or to culture it within suitable cells.

In addition to its inability to stimulate the immune system, the scrapie agent has been found to display several other unusual characteristics.[13] Firstly, it seems to be much more resistant than conventional viruses to degradation by chemicals, radiation, and enzymes that normally destroy nucleic acids. That last feature has prompted some scientists to conclude that it may not contain nucleic acid at all! Secondly, several studies have suggested that the scrapie agent is extremely small – much smaller than a conventional virus and perhaps not much bigger than an individual protein molecule. Almost all of these unconventional claims about the scrapie agent, however, are challenged by the results of other studies that suggest it may turn out to be a fairly conventional virus after all.[14] Overall then, scrapie studies are in a state of considerable confusion which might not be resolved for quite some time.

Scientists interested in scrapie and other similar diseases are currently in much the same position as Ivanovski and Beijerinck were at the turn of the century. They have on their hands infectious agents that appear to be different from those already known, and whose nature can only be the subject of fairly wild speculation. Rather than looking at the twists and turns of the evidence in any detail, I would simply like to outline some of the main proposals put forward to explain the properties of the scrapie agent. One of these strange proposals (summarised in figure 13.1) might eventually turn out to be true.

The most radical suggestion is that the scrapie agent is simply an "infectious protein" that contains no nucleic acid at all. In order to be infectious, of course, such a protein would have to be able to multiply. So how could a protein that passes from cell to cell without any accompanying nucleic acid manage to multiply? Some scientists, completely abandoning the conventional wisdom about cell chemistry, have suggested that the infectious protein might be capable of self-replication. In some way, for example, the fully formed protein might be able to act as a template on which the amino acids of a "daughter" protein could be linked up in the correct sequence (see figure 13.1a). Such daughter proteins might be identical to their parent molecule, or they might be "complementary" (just like complementary nucleic acid chains) and so able to themselves act as templates on which an identical copy of the parent protein would be produced. Alternatively, self-replicating infectious proteins might be able to initiate some kind of "reverse translation" process, giving rise to a nucleic acid that would then be used to manufacture more copies of the protein in the conventional way (see figure 13.1b).

Confirmation of either of these proposals for protein self-

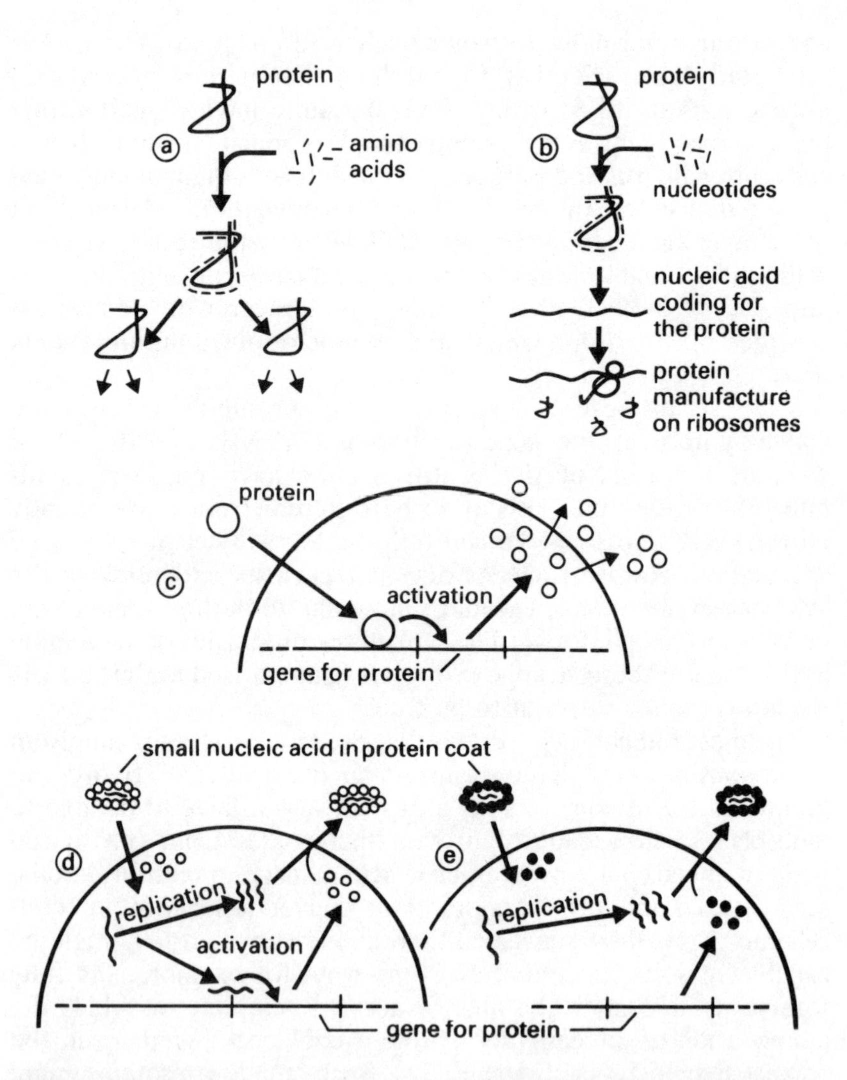

Figure 13.1 Some of the novel types of infectious agents proposed as the cause of scrapie. a) infectious protein that acts directly as the template for its own replication; b) infectious protein that can somehow bring about the manufacture of a nucleic acid that codes for it; c) "escaped" cell protein that can enter other cells and switch on the gene that codes for it; d) small nucleic acid that can switch on the gene for a cell protein that can bind to and protect it;
e) small nucleic acid that can become encased in already available cell protein

replication would be a truly revolutionary discovery, shattering at least one central dogma of modern molecular biology. Most scientists prefer to dismiss such "unlikely" possibilities and turn instead to more "reasonable" explanations for the multiplication of infectious proteins – if such proteins do in fact exist.

One feasible possibility is that infectious proteins might be proteins that have somehow managed to escape from the cells that make them. They might, for example, escape from one cell, enter into others, and then switch on the gene within the newly "infected" cells that codes for the infectious protein itself (figure 13.1c). So first of all by escaping from cells that contain them, and secondly by being able to switch on the gene that encodes them, such infectious proteins could spread throughout any population of cells containing the necessary gene. Such escaped proteins would, of course, not really be "foreign invaders", simply host cell proteins that had got out of control. This would neatly explain why the scrapie agent does not stimulate the immune system, since the immune system would not recognise it as foreign.

All of the radical "infectious protein" proposals, however, stem from the assumption that the scrapie agent does not contain any nucleic acid. Some scientists challenge this assumption, suggesting that the resistance of scrapie to treatments expected to destroy nucleic acids might simply be due to an unusually effective protective protein coat. But even if the scrapie agent does turn out to contain nucleic acid, other evidence suggests that there cannot be very much of it – certainly not enough to serve as the gene(s) needed to encode the agent's protein coat. This has led to various other novel ideas to explain how infectious agents containing just a tiny piece of nucleic acid might be able to multiply.

It has been suggested that the scrapie agent's nucleic acid (if it has any) might somehow be able to switch on the gene(s) (again part of the host cell genome) that codes for the scrapie agent's protein (see figure 13.1d). The nucleic acid would itself be replicated by host cell enzymes, allowing newly made scrapie particles to be assembled and released just like a conventional virus. Again, the scrapie protein would be native to the host cell and so unable to stimulate the host organism's immune system.

A final possibility is that the scrapie agent is essentially a small infectious nucleic acid molecule which can enter cells, be multiplied by host cell enzymes, and bind to host cell proteins prior to its release (see figure 13.1e). In this scheme the scrapie nucleic acid does not need to be able to switch on any host cell genes, it simply surrounds itself with proteins normally present in the infected cell. As before, the fact that the scrapie-associated proteins would really belong to the host cell would explain why they do not act as effective antigens.

So these are just some of the weird explanations that scientists are turning to in the attempt to explain the unusual characteristics of the infectious agent responsible for scrapie (and by implication the similar agents that cause Jakob–Creutzfeld disease and perhaps other still mysterious diseases of humans). If any of these unusual infectious agents really do exist then obviously, while resembling viruses in various ways, they are not true viruses. Various names for them have been proposed, including "prions" (for "proteinaceous infectious particles") and "virinos". But of course settling on names is much less important than determining the true nature of the agents involved.

Amidst all the speculation about novel infectious agents that may cause disease in humans and other animals, a class of unconventional virus-like agents has been conclusively shown to cause certain diseases of plants. From the early 1970s onwards it has become apparent that several damaging diseases of crop plants are caused not by viruses, but by tiny naked RNA molecules known as "viroids".[15] Now having been told that they are infectious naked RNAs, you might be tempted to think that viroids behave simply like viruses that manage to do without their protein coats (I told you in chapter 3, after all, that the naked RNA of tobacco mosaic virus can cause a normal infection when introduced into suitable cells); but this does not seem to be the case. Apparently the RNA of a viroid *does not code for any protein molecules*, so every step in viroid multiplication must be carried out by enzymes native to the infected cell.

Viroids, it seems, are just fragments of RNA that can be taken up by cells, replicated by the cells' own enzymes, and then released from the cells to initiate the same cycle of viroid multiplication elsewhere. In many cases the presence of viroids apparently does the infected cells no harm, but in a few cases the damage can be considerable. The tiny viroid shown in figure 13.2, for example, can cause potato tubers to become cracked, gnarled and elongated. Other viroids can devastate coconut tree plantations and damage several other important crops such as hops, cucumbers and avocados. The economic losses caused by these tiny and

–CGGAACUAAACUCGUGGUUCCUGUGGUUCACACCUGAC
CUCCUGAGCAGAAAAGAAAAAAGAAGGCGGCUCGGAGGA
GCGCUUCAGGGAUCCCCGGGGAAACCUGGAGCGAACUGG
CAAAAAAGGACGGUGGGGAGUGCCCAGCGGCCGACAGGA
GUAAUUCCCGCCGAAACAGGGUUUUCACCCUUCCUUUCU
UCGGGUGUCCUUCCUCGCGCCCGCAGGACCACCCCUCGC
CCCCUUUGCGCUGUCGCUUCGGCUACUACCCGGUGGAAA
CAACUGAAGCUCCCGAGAACCGCUUUUUCUCUAUCUUAC
UUGCUUCGGGGCGAGGGUGUUUAGCCCUUGGAACCGCAG
UUGGUUCCU–

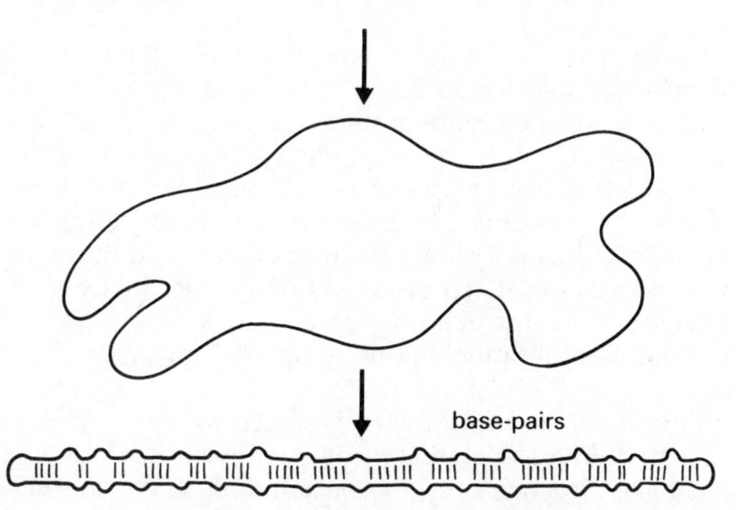

Figure 13.2 The structure of viroids, the smallest known agents of infectious disease. The viroid known as "Potato Spindle- Tuber Viroid" is a tiny single-stranded RNA molecule made up of just 359 nucleotides linked in the precise sequence shown above. Normally the ends of the RNA are joined to make an RNA circle, which then forms extensive base-pairing to adopt a linear rod-like shape overall. Other viroids have a very similar overall structure

absurdly simple infectious agents can also be considerable.

Two main questions are raised by the discovery of the viroids: Firstly, can they infect and cause disease in humans and other animals as well as plants? Secondly, how do such simple agents manage to cause disease in the first place? The first question is, of course, of great importance to the doctors and scientists grappling with the many still mysterious and apparently infectious diseases known to afflict mankind. Might viroids be responsible in some cases? The only answer that can be given at the moment is that there is no really firm evidence suggesting that viroids infect humans or other animals, but there seems to be no good reason in principle why they should not be able to attack us. The search for viroids that infect humans will certainly continue.

Of course the existence of infectious agents similar to the viroids, but made of naked *DNA*, is another possibility to bear in mind. Indeed, one group of scientists have already claimed to have identified a "DNA-viroid" that can cause cancer in hamsters.[16] For the time being, however, other scientists remain sceptical of such reports.

The second major mystery about the viroids (and any naked DNA relatives they might have) concerns how they manage to cause disease. The currently available answer is simple – nobody knows. This lack of knowledge, however, has not prevented various scientists from putting forward some plausible proposals. Viroids may, for example, be able to bind by base-pairing to some of the genetic material within an infected cell, and thus interfere with the activity of cellular genes and RNA. Alternatively, viroids may damage cells just by being there, keeping cell enzymes busy with viroid multiplication and using up vital raw materials in the process.

So the overall message of what has been written in this chapter so far is that there are a great many mysteries still to be solved in the world of virology; not just concerning how the conventional viruses work and cause disease, but also concerning various strange virus-like agents that may share some characteristics with the viruses while also displaying unusual properties all of their own. The final mystery that I want to consider is not only one of the most intriguing, but also one of the most fundamental: where do the viruses and other virus-like agents come from? How did they first arise and do the similarities between the structure of viruses and such novel agents as the viroids reflect a common evolutionary past? In other words, what do we know about viral *origins*?

Origins[17]

Where have the viruses come from? Where, indeed, *do* they come from, supposing that novel viruses are being created to this day?

The simplest answer is firstly, that nobody knows; and secondly, that it is quite possible that different types of virus have arisen in completely different ways. All I want to do here is outline some of the most plausible possibilities and look at the major strands of evidence backing them up.

There have always been three main theories advanced to explain the origin of the viruses. Firstly, they might have originated very early in evolution before the advent of complex cellular life. In other words, modern viruses might be the descendants of primitive early forms of "life" which consisted of little more than self-replicating nucleic acids floating free in some sort of primordial "soup". Secondly, they might be derived from ancient parasitic cells which invaded other cells and then gradually became simpler and simpler – relying increasingly on their host cells to oversee their survival and multiplication. Thirdly, the viruses might have evolved from the genetic material of cellular life, little bits of which might be able to "escape" to take up the independent existence of a virus.

Competing explanations for such impenetrable mysteries as the ancient origins of modern life-forms tend to come in and out of fashion according to the interests and prejudices of the scientists of the day. The present status of the three tales of viral origin summarised above is as follows: the first idea, that the viruses might be derived from free-living pre-cellular organisms, is generally dismissed as unlikely. The second story, involving the gradual simplification of parasitic cells as they become ever more dependent on their hosts, is occasionally invoked to explain the origin of some complex viruses. This third theory, proposing that viruses have evolved from escaped cellular genetic material, is currently very much the vogue. This third idea owes much of its dominance to recent discoveries in molecular biology, which have revealed suggestive similarities between some viral genomes and some forms of free genetic material found in all cells. Since it is the dominant theory of today, this "escaped genes" hypothesis will be examined first and in most detail (see figure 13.3).

In general, the fact that many viruses can integrate their genetic material *into* their host cells' genomes, makes it plausible that viral genes might be derived *from* the host cell genomes in the first place. Many different specific routes by which cellular genetic material might escape and evolve into a virus have been proposed. Rather than going into all the details of schemes that are still

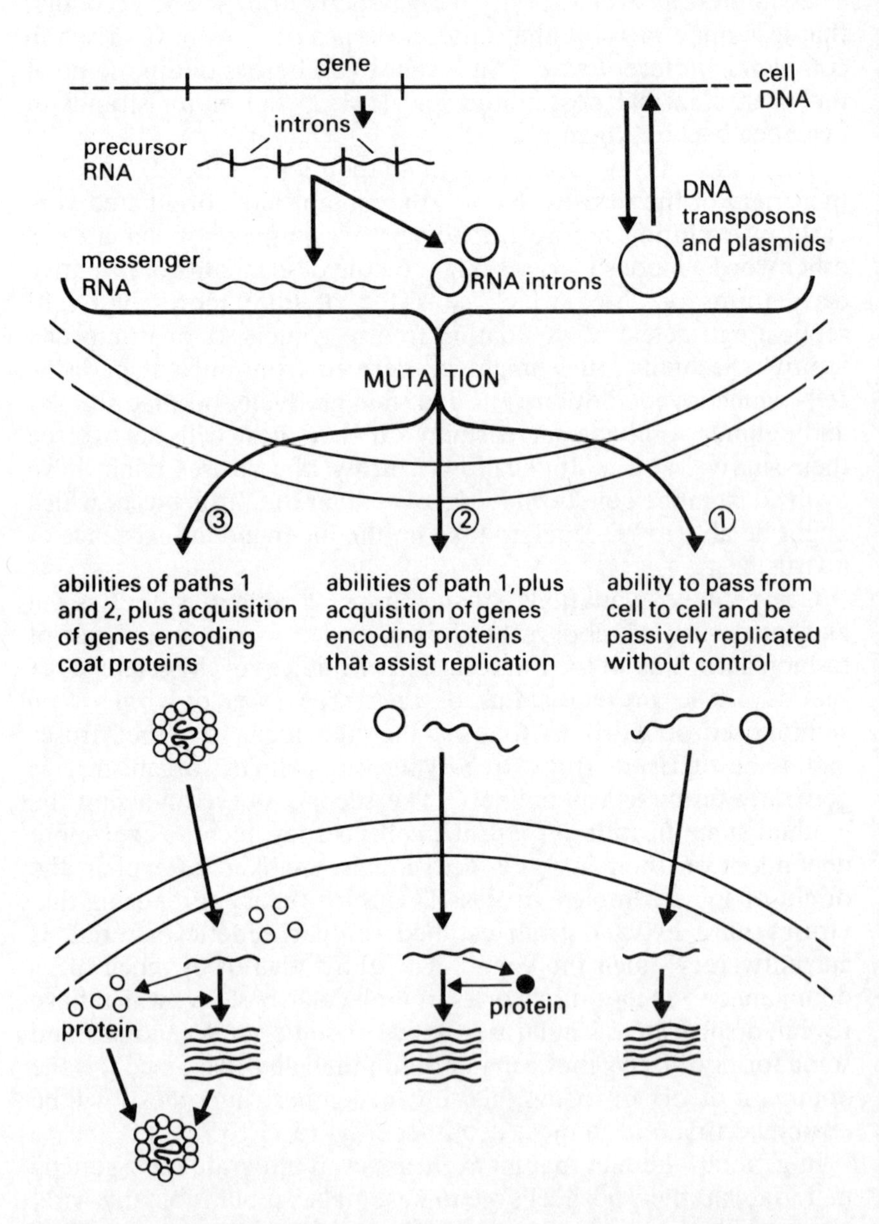

Figure 13.3 Possible pathways for the origin of viruses and viroids from cellular genetic material (see text for details)

entirely hypothetical, I will simply outline the major steps that might be involved overall.

Firstly, of course, suitable pieces of genetic material must arise free within a cell, detached from the main bulk of the cell's chromosomal DNA. If we allow this genetic material to be either RNA or DNA, then there are several possible candidates. The most obvious one is simply messenger RNA, which *must* be released into the cell at large in order to perform its normal function; but there are at least two rather more attractive possibilities.

Over the past decade it has become apparent that most of the genes in higher organisms are "split", in the sense that the regions of a gene that code for protein are separated by seemingly meaningless sections of DNA known as "introns". During the transcription of a split gene the introns are copied into RNA as well as the protein-coding sections of a gene, yielding a long precursor RNA in which sections of the eventual messenger RNA are separated by RNA copies of the introns (see figure 13.3). To produce mature messenger RNA the introns must be cut out from this precursor and the protein-coding sections joined together (a process known appropriately as "splicing"). In addition to making mature messenger RNA, splicing releases free introns into the cell, probably in the form of small RNA circles. These released RNA introns provide us with a second class of free cellular nucleic acid from which some viruses might evolve.

The third favoured candidates for the role of cell-derived precursors of viral evolution are made out of DNA, not RNA. The genomes of both bacteria and higher organisms contain small sections of DNA that can spend much or part of their time as free DNA circles, independent of the cellular chromosomes. Bacteria, for example, contain circles of DNA known as "plasmids" which spend most of their time in the free state, but which can sometimes integrate into the bacterial chromosome, behave like normal chromosomal DNA for a while, and then detach themselves and once more take up an independent existence. The genomes of both bacteria and higher organisms contain short sections of DNA called "transposons" that can *occasionally* move from place to place on the cell chromosomes, sometimes existing during their wanderings as free circles of DNA. Such transposons can integrate into a new site on a chromosome; and then later either be used as a template to make another free copy for transfer to yet another site, or else the original DNA might detach itself from its temporary resting place to integrate again somewhere else. In many cases transposons may spread copies of themselves from site to site without an intervening stage in which they exist as free DNA; but whatever the precise details of transposon spread are (many of these details still being far from clear), it is the stage during which they can be free in the cell that we are interested in here; for it is during that stage that they might serve as precursors of the viruses.

So there are at least three classes of genetic material available within cells to serve as the raw materials for conversion into viruses. What changes would be needed to send messenger RNA, RNA introns or DNA plasmids and transposons down the evolutionary path towards the viruses? The simplest possibilities require very little change at all – just the chance escape of a piece of genetic material from its parent cell, and the ability of the escapee to enter into and be replicated within other cells (perhaps of a different type). If the structure of the escaped genetic material (probably modified by appropriate mutations) allowed it to multiply out of control and swamp the "infected" cells, then these cells might disintegrate and so release many more copies of the "infectious" genetic material which would then repeat the cycle in other cells. Overall, then, according to this scenario any piece of cellular genetic material whose replication got out of control, and which was able to escape from cells, survive for a while and then be taken up by other cells, might quickly begin to behave more like a virus than a piece of cellular nucleic acid.

The life-cycle of the viroids is very similar to the sort of process outlined above, and examinations of viroid structure have yielded fascinating clues supporting the involvement of such a sequence of events in viroid evolution. For example, the RNA of many viroids includes sections whose nucleotide sequences are complementary to important regions at the boundaries of RNA introns.[18] So could viroids have originated when some (mutant?) RNA introns were copied into complementary RNAs, which then escaped from their cells to slowly evolve into the viroids we find in cells today? This is certainly a fascinating idea, and David Zimmern of Cambridge University has suggested that the natural activities of introns might encourage such an intron into viroid change.

Zimmern has proposed that introns might normally be replicated after splicing and then passed from cell to cell, allowing one cell to "know" which genes are being spliced (and are therefore active) in its neighbours. Introns might thus act as intercellular messengers, helping neighbouring cells to work in harmony rather than in opposition.[19] Although Zimmern's idea is still purely speculative it might, if true, go a long way towards explaining why viroids can successfully multiply and move from cell to cell – because they may have evolved from introns whose proper role in life is to do just that.

Whatever its relevance to viroid evolution, the discovery that sections of viroid RNA are complementary to sections of intron RNA offers one possible explanation for the damaging effects that viroids can have on infected cells: Perhaps they interfere with RNA splicing by binding to important regions of some RNA introns.

Another intriguing discovery about viroid structure is that the nucleotide sequence of some viroids is very similar to sections of some transposons and retroviruses, leading to the suggestion that these entities might also be related to the viroids in some way.[20] The mystery of viroid origins is certainly far from solved.

To many people the idea that free cellular DNA or RNA molecules might be able to become infectious while changing very little from their original state seems rather far-fetched. Much more plausible, they feel, would be the occasional occurrence of a transposon or plasmid that had not only mutated to a form that could replicate out of control, but which had also acquired a gene encoding a protein that actually catalysed that replication. Mutant transposons and plasmids might thus be able to become simple naked "viruses", able to pass from cell to cell and make the enzyme(s) needed to ensure their further multiplication and spread. Evolution, driven by simple mutations and the occasional incorporation and modification of appropriate cellular genes, might slowly provide the gene complement of a fully-fledged virus; including the genes needed to make proteins that bind to the genetic material to serve as a viral coat. So with time transposons or plasmids "belonging" to cells would change into viruses capable of infecting and perhaps destroying the cells.

Such a series of events might also be able to convert introns or messenger RNAs into viruses, but several factors make transposons and plasmids the best candidates. Some plasmids, for example, can certainly pick up cellular genes which might serve as the raw material for evolution into viral genes, and similar processes by which transposons might be able to collect new genes from the cell have also been proposed. As far as the evolution of coat proteins is concerned, some plasmids encode proteins that form a tube connecting two bacterial cells, and through which the plasmid DNA can pass from one cell to the other. It might not take too many mutations to convert such plasmid proteins into the protective proteins of a viral coat.

Striking similarities have been found between the nucleotide sequences of retroviruses and transposons, particularly in the sections of nucleic acid likely to be crucial for the integration of retroviral or transposon DNA into the DNA of a cell.[17] These similarities have led to the suggestions first of all that retroviruses might have evolved from transposons (and perhaps vice-versa in some cases); and secondly that other types of virus carrying both RNA or DNA genomes might then have evolved from the retroviruses. When constructing models of viral evolution we need not worry too much about converting an RNA virus into a DNA virus or vice-versa; after all the life-cycle of the retroviruses clearly demonstrates that viral genomes might easily be converted from one form of nucleic acid into the other.

Overall, then, the "escaped genes" story of viral origins has a lot to support it, and although the details of the evolution of any particular virus by this route might often be fairly complex, the simplicity of the idea overall makes good sense: Viruses might in general arise when sections of cellular nucleic acid mutate, begin to be replicated out of control, and then slowly evolve the ability to encode various proteins that help the fledgling viruses to successfully pass from cell to cell and multiply. Although most of the viruses around today are probably of very ancient lineage, if the escaped-genes hypothesis is true it means that new viruses might be slowly evolving from cellular genetic material all the time, right up to the present day. Indeed, some entities have already been found within modern cells that behave a bit like viruses and a bit like plasmids or transposons – making it difficult to decide which is the most appropriate label.

As more and more is discovered about the various types of genetic material found within cells, a grey area is emerging between truly chromosomal genes and the "foreign" genes of viruses. Plasmids, transposons and perhaps other as yet undiscovered and *partially* independent genetic entities may be the intermediate links in a continuous chain between the genes of a cell and the genes of the viruses that infect cells. While viruses generally harm cells or have no effect on them, and cellular genes are clearly of benefit to cell survival, the emerging intermediates such as transposons may sometimes be harmful, sometimes beneficial and sometimes a bit of both.

Despite the dominance of the escaped-genes theory of viral origins, the alternative idea that viruses might have evolved by the increasing simplification of cellular parasites still looks attractive to some people in some cases. The large and very complex poxviruses probably provide the best example. Poxviral genomes are about the same size as the genomes of the smallest cells, and the virus particles themselves are also about as big as the smallest cells.

Some scientists find it difficult to see how viruses with so many genes could have evolved from small pieces of escaped genetic material which then proceeded to accumulate more and more genes. As evidence to back up their scepticism they cite various studies suggesting that evolutionary pressures seem to encourage medium-sized viruses to mutate into *smaller and simpler* forms, rather than change towards more complex structures such as the poxviruses.[17] So rather than turning to the escaped-genes theory to explain the origin of the poxviruses and some similar complex viruses, these scientists suggest that the sorts of events summarised in figure 13.4 might have been involved.

First of all, very simple bacteria or similar primitive organisms would invade the cytoplasm of more complex cells and become parasites of these cells. The parasitic cells would initially just take up some simple raw materials from their host cells' cytoplasms – generally using the host cells as comfortable "homes" provided with many of their most basic needs. Through the generations mutant parasites might then arise either lacking outer membranes and walls completely, or else making very "leaky" ones. This would make the parasitic genetic material accessible to the enzymes of the host cells, which might then perform many of the tasks that previously had to be performed by the parasites' own enzymes. Most importantly, the parasites' messenger RNAs would be able to be translated by the protein-making machinery (ribosomes, transfer RNA, etc.) of the host cells. Gradually, the genes coding for proteins and RNAs whose functions were duplicated by proteins and RNAs available in the host cells, would be lost from the parasitic genetic material. The parasites would cease to make outer membranes and walls but would mutate to produce proteins that could surround and protect the parasitic

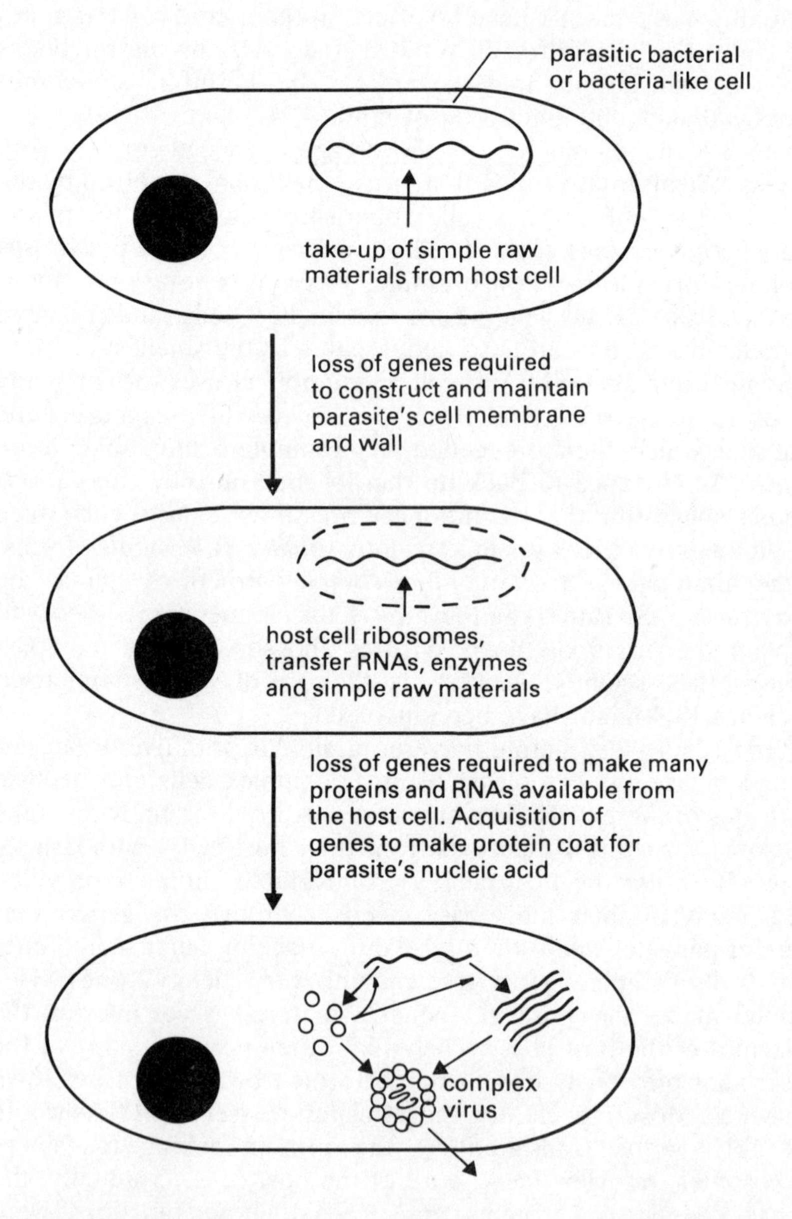

Figure 13.4 Possible key steps on the way from parasitic cells to complex viruses

genome during its passage from cell to cell. So with time, and thanks to the accumulation of appropriate mutations, the simple parasitic cells would change into large and complex viruses. The process of simplification, with more genes being lost or made inactive, might be continuing within some poxviruses and other similar viruses right up to the present day.

So both of the theories of viral origin discussed so far have their attractions, and they each might be correct in different individual cases. I should repeat and emphasise, however, that the escaped-genes idea is currently by far the most favoured.

What about the very first idea mentioned at the start of this section – the possibility that modern viruses are the descendants of primitive pre-cellular forms of life? I have already said that this idea is largely dismissed as highly unlikely. The justifications for this dismisal usually go something like this: in almost every detail the biochemical activities of the viruses seem to be identical to those of their host cells. For example, viruses use exactly the same genetic code as the cells they infect; their genetic material is replicated, transcribed, translated and generally processed in much the same overall manner as the genetic material of the infected cells; the various amino acids are found in viral proteins with much the same frequency as they are found in the proteins of cellular life – and so on. So in general, while viruses are certainly much *simpler* than the cells they infect, there is little evidence to suggest that they are more *primitive*.

Such arguments cannot *prove* that no modern viruses are derived from primitive free-living nucleic acids that evolved *in harmony* with cellular life – slowly gathering mutations that ensured the viral nucleic acids were efficiently replicated, transcribed and translated by the systems being developed to perform these tasks on cellular genetic material. Of all the possible tales of viral origin this one is the least open to investigation. After all, the events concerned would have taken place many millions of years ago; they will probably not be paralleled by anything going on in the biosphere today, and we can hardly go back to these ancient times to take a look!

The impossibility of ever really *knowing* what happened in the past to bring the modern viruses into being (no matter how plausible our theories might seem to be) makes the mystery of viral origins the toughest of them all. Scientists investigating any of the other mysteries discussed in this chapter know that, no matter how intractable their problems might be, the answers lie right before them within the viruses, cells and tissues being studied on the laboratory bench. But even if in a few cases important changes on the path from, say, transposons to viruses are observed within modern cells, we will never really know which of the present-day fully fledged viruses actually developed in similar ways.

So on that rather frustrating note of harsh reality I want to leave the subject of viral origins and also the many other viral mysteries that still remain to be solved. The final topic for consideration in this rather rapid journey through the world of the viruses is one of great hope and practicality – the increasingly successful attempts to turn the viruses from agents of pestilence and misery into valuable biological systems that we can *exploit* for our own benefit.

CHAPTER 14

Exploitation – from menace to boon?

The story of mankind's relationship with the viruses has for millennia been a one-sided tale of death and disease. Smallpox, influenza, yellow fever, rabies, polio and even cancer – the list of illnesses caused by the viruses is long and the tally of victims numbers many many millions. Of course in addition to afflicting humans, the viruses have also caused great damage through the ages to both our livestock and our crops. Things began to change for the better a little less than 200 years ago, when Edward Jenner sent us down the road towards safe and effective anti-viral vaccines. In a few decades, with just a little more research and a great deal of political will and organisational effort, vaccination could be protecting the entire population of the world against most of the really serious viral diseases. Also, an impressive battery of anti-viral therapies might well be available to combat those infections that slip through the vaccination net. The generations of humans alive today are witnessing a complete transformation in mankind's relationship with the viruses – the elderly having seen some of the first blows struck against the viral threat, while those newborn today may be in on the final victory. But the turning of the tables on the viruses is not going to stop simply with their defeat or control. Instead, it is becoming clear that the viruses may become useful allies in our unending efforts to manipulate the world around and within us. After millennia during which they only harmed us, the viruses are now being made to *help* us.

In their efforts to press the viruses into the service of their own ends, scientists are finding jobs for unchanged viruses gathered from nature, novel viruses created by the tricks of genetic engineering, and purified viral components such as protein coats and membranes. Before turning to the possibilities offered by more modern techniques, however, it is appropriate to begin our examination of the exploitation of viruses with an old idea that was tried for a while, abandoned, and is now once again showing some promise – the use of bacteriophages as "antibiotics" to combat dangerous bacterial infections.

Viral "antibiotics"?

As I briefly mentioned in chapter 1, the possibility of using bacteriophages to treat bacterial infections was one of the first things to occur to the discoverers of the bacteriophages, such as Frederick Twort. But early attempts to make this good idea yield useful results were disappointing and were eventually abandoned. More recent attempts have been much more promising, perhaps thanks to improved methods for the culture and purification of suitable bacteriophages. Seventy years or so after it was first suggested, the use of bacteriophages to fight infection might be about to flourish into routine medical practice.

Some of the most dramatic successes have come from Poland, thanks to the efforts of Stefan Slopek and his colleagues at the Polish Academy of Sciences' Institute of Immunology and Experimental Therapy. For example, they recently selected 138 patients suffering from long-term and drug-resistant bacterial infections, and treated them with bacteriophages chosen as the most virulent against the bacteria concerned. The viruses were either swallowed along with a glass of water, or administered directly to open wounds. In 121 of the patients the infections were completely overcome, while all of the remaining 17 showed at least some improvement.[1] These are truly remarkable results, especially when you consider that in most cases Slopek's team turned to the bacteriophages as a last resort.

The particular infections that the bacteriophages were shown to be effective against included furunculosis (widespread festering boils), bronchitis, pneumonia, chronic pharyngitis, dysentery and infections of the urinary tract; and remember, in almost all cases the infections had proved completely resistant to modern antibiotics.

Slopek's achievements are only part of a more widespread resurgence of interest in the power of bacteriophages to combat bacterial infection. British scientists, for example, have recently shown that bacteriophages can also be used to successfully treat intestinal infections in piglets, calves and lambs.[2] These modern successes should ensure an intensifying effort throughout the late 1980s to find out just how useful the anti-bacterial activities of bacteriophages might be – especially in view of the increasing problems being posed by the emergence of many strains of drug-resistant bacteria.

In many ways bacteriophages look much more attractive anti-bacterial agents than some modern drugs. They are highly specific (each bacteriophage attacking only particular types of bacteria), so they can be selected to leave potentially beneficial bacteria found within the body unharmed. They apparently cause side-effects only rarely, and fairly trivial ones at that, and they are effective in small doses thanks to their ability to multiply within the bacteria they eventually destroy. All these advantages, in addition to the cheapness and ease of bacteriophage production, have led some people to suggest that a second anti-bacterial revolution might well be on the way. But others are much more sceptical and cautious, pointing out that bacteria *resistant* to any bacteriophages we deploy against them are at least as likely to arise as bacteria resistant to various drugs. Nevertheless, if used sensibly and selectively, bacteriophages might at the very least soon have a small but significant role to play in fighting off bacterial infections against which modern drugs are currently useless. The discovery that viruses really can be used to *treat* infectious diseases, rather than to cause them, certainly adds a new class of weapon to the armory of modern anti-microbial medicine.

The conquest of bacteria is not the only use that scientists are finding for the viruses gathered ready-made from nature. Various research teams throughout the world, for example, have used the glandular fever-causing Epstein–Barr virus (which can also occasionally cause cancer) to "transform" selected antibody-producing B-cells. The transformed B-cells can then survive and multiply indefinitely, producing large quantities of specific antibodies which might be useful in both medicine and industry.[3] So the properties of a virus that allow it to cause cancer can be exploited to provide us with supplies of natural proteins that can be purified and put to good use.

Having looked at just two examples of the uses being found for naturally occurring viruses, I want to move on to consider some uses of the reconstructed viruses that can now be made by genetic engineering.

We have already seen that genetic engineers are changing some viruses into novel forms for use as live vaccines. The genes for influenza and hepatitis B virus proteins, for example, are being stitched into the vaccinia virus genome to produce "engineered" vaccinia viruses that will hopefully be able to vaccinate us against hepatitis B and the flu. This approach will probably not be restricted to the development of new *anti-viral* vaccines. Genes encoding proteins belonging to other types of parasite, such as the protozoan that causes malaria, are also being added to suitable viral genomes.[4] So in general the viruses are being turned into versatile "ferrymen" which vaccine designers can use to carry a whole range of foreign antigens into humans and our livestock.

Another ferrying task which the viruses are being persuaded to perform for us is the actual transfer of new *genes* into various types of recipient cells. Bacteriophages, for example, are now routinely used by genetic engineers to transfer foreign genes into bacterial cells. The gene coding for any wanted protein can be linked up into the genetic material of a bacteriophage; and then when the virus infects suitable bacterial cells it will carry the foreign gene in with it. Once inside a bacterium, the transferred gene may then begin to direct the manufacture of plentiful supplies of the protein it codes for.[5]

The most exciting possibilities offered by the development of viruses that can carry genes into chosen cells, however, concern the transfer of new genes into *human* cells rather than into bacteria. Richard Mulligan and other scientists at the Massachusetts Institute of Technology, for example, are busy trying to construct retroviruses that will carry new genes into human cells and so perhaps be able to cure some serious genetic diseases.[6]

"Gene therapy"

Many human diseases are caused by the lack of, or defects in, certain crucial genes. Serious and potentially fatal anaemias such as "sickle cell anaemia" and "thalassaemia" are caused by the lack of normal genes encoding the proteins of haemoglobin – a multi-subunit protein responsible for carrying oxygen around the body in red blood cells. "Adenosine deaminase deficiency", caused by lack of a normal gene encoding the enzyme adenosine deaminase, leads to a crippling deficiency of the immune system. In "Lesch-Nyhan syndrome" the gene for a protein known by the grand title of hypoxanthine-guanine phosphoribosyl transferase is missing, leading to mental retardation, cerebral palsy and severe gout. And so the list goes on. The aim of scientists like Mulligan is to exploit retroviruses to take the needed genes into the cells of patients suffering from such genetic diseases, integrating the genes into the patients' own DNA to allow the diseases to be permanently overcome.

Many of the diseases that might be amenable to "gene therapy" involve genes that are normally active in the cells of the blood. All of these cells are derived from a small population of constantly dividing bone marrow cells known as "stem cells", so it is into the stem cells that the genes required to cure the diseases must be put. Mulligan and his colleagues are using genetic engineering technology to insert the genes for human haemoglobin proteins into suitable retroviral genomes. The engineered retroviruses are then being used to infect the bone marrow cells of mice suffering from a genetic defect of haemoglobin production. The hope is obviously that the retroviruses will carry a suitable haemoglobin gene into the mouse stem cells, where it will become a permanent integrated part of the stem cell DNA, being replicated and passed on during cell division, to eventually appear in mature blood cells in which it will code for the protein needed to cure the disease. Success in using such methods to cure genetic disease in mice will clear the way for attempts to cure human diseases in a similar way (see figure 14.1).

As I write, no-one has yet cured a laboratory animal's genetic disease by using viruses to carry the required genes into appropriate cells, but preliminary experiments have suggested that the prospects for eventual success may be quite good. The Massachusetts Institute of Technology team, for example, have successfully used an engineered retrovirus to incorporate a bacterial gene into the genetic material of the stem cells of mice. Success with a useful gene needed to cure genetic disease will hopefully be the next step.

Getting a retrovirus to insert a therapeutic gene *into* the DNA of cells deficient in the gene is unfortunately only the first part of the problem; for even if the virus successfully makes the new gene a permanent part of a cell's genome, there is no guarantee that the gene will *work properly* in its new environment. Many of the

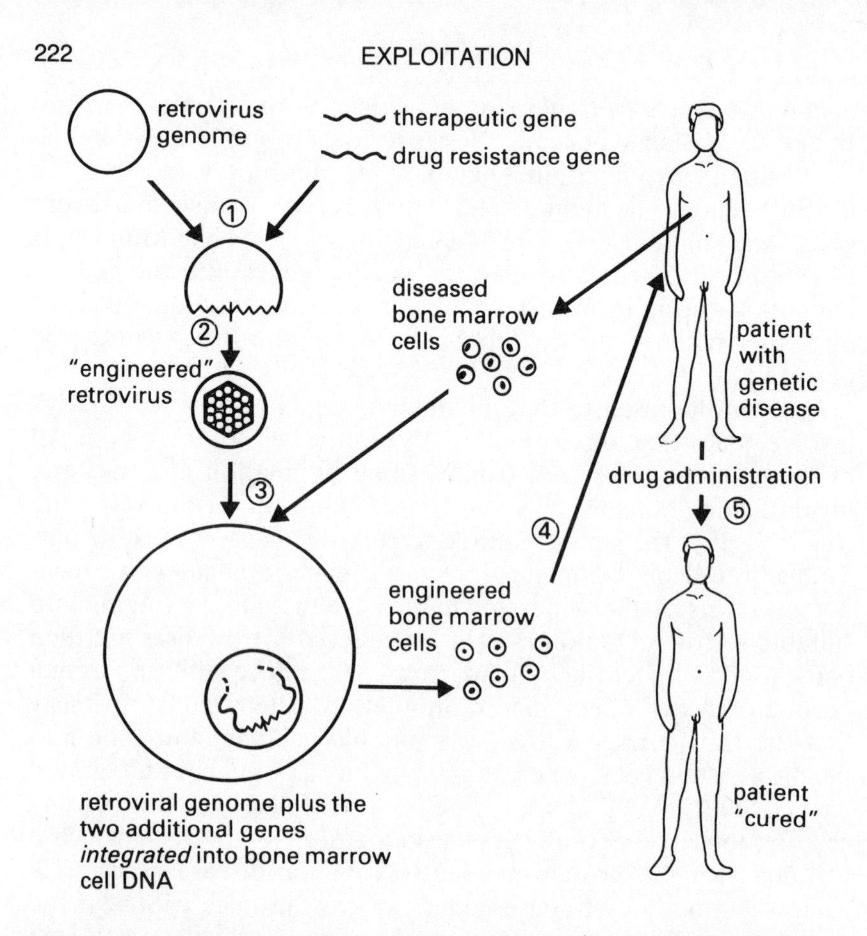

Figure 14.1 One way in which retroviruses might be used to cure genetic diseases affecting blood cells. 1) the gene required to correct a disorder is added to a retroviral genome, along with a gene conferring resistance to a toxic drug; 2) retrovirus particles containing the engineered genome are formed within cultured cells and harvested; 3) cells taken from the patient's bone marrow are infected with the retrovirus and then; 4) returned to the patient; 5) the patient is treated with the drug to which the engineered cells will be resistant, allowing engineered stem cells to multiply, come to dominate the stem cell population, and produce blood cells containing the active therapeutic gene. Drug administration could then be stopped

factors involved in *controlling* gene activity – determining where and when a gene is actively decoded into protein and what amounts of protein are made – are still unknown. So in some ways scientists do not even know what they should be *trying* to do to get artificially inserted genes to be properly expressed in humans or laboratory animals, quite apart from knowing how they might actually be able to do it. However, the problems associated with getting transferred genes to work properly are unlikely to be insurmountable. Knowledge about the control of gene activity is increasing all the time; and some initial successes are already indicating that the appropriate expression of the genes we put into animals and humans might eventually be quite readily achieved.[7]

I for one will be very surprised if genetic diseases are not being cured, or at least considerably alleviated, within my lifetime by the insertion of new genes into the cells that need them. Stitching the genes into the genetic material of retroviruses could well be the way in which they will be put in. But using engineered retroviruses is certainly not the only way to get new genes into the DNA of human and other animal cells. Various other types of virus are also being altered to act as the necessary ferrymen,[8] while techniques that do not rely on viruses at all have also proved quite effective. Foreign genes can integrate into cellular DNA after being injected directly into the recipient cells' nuclei, for example; or genes simply added to the medium around cultured cells can, under appropriate conditions, be taken up into the cells to become spontaneously integrated into the cellular DNA.[9] But retroviruses might well turn out to offer the most efficient way to get the job done successfully; firstly because they are obviously adept at binding to and invading cells, and secondly because the retroviral life-cycle ensures that any foreign genes they carry with them will become integrated into cellular DNA.

Obviously the scientists trying to cure genetic diseases by using retroviruses to ferry in the appropriate genes will need to be careful to ensure that the viruses cannot *cause* disease rather than cure it. Some retroviruses, after all, can cause cancer, so the need for care is obvious. This is unlikely to be much of a problem, however, as it is quite easy to obtain or construct suitable defective or altered retroviruses that will carry genes into the DNA of cells you want to receive them, but do little else.

No matter how successful scientists are in using retroviruses (or any other gene transfer systems) to cure genetic diseases, the techniques considered above are not going to rid us of all genetic disease. For one thing, most genetic diseases are caused by *too much* genetic material, rather than by the lack of any particular genes; so to cure these diseases ways will need to be found to *remove or destroy* genes rather than add any new ones. Secondly, curing genetic disease by inserting new genes into human cells only looks feasible if the cells concerned belong to a population that is constantly replacing itself through cell division and cell death. This is essential because it is only possible to add a new gene to a small percentage of the cells in any cell population. You must then rely on the rapid multiplication of the engineered cells to allow them to come to dominate the population overall. Obviously, to ensure such dominance you must in some way give the engineered cells some advantage over unaltered ones, allowing them to multiply much more quickly.

This necessity for cell division in the affected cell population is what makes diseases of blood cells so attractive to the genetic engineers, because all blood cells are derived from the actively dividing stem cells of the bone marrow. One way to achieve the dominance of engineered cells within a patient's bone marrow might be to insert a gene that will make the engineered cells resistant to some toxic drug, along with the gene needed to cure a disease. Administering appropriate dosages of the drug to the patients might then ensure that those cells carrying the drug resistance gene, and therefore also the wanted therapeutic gene, will be able to survive and multiply much more effectively than the unaltered cells (see figure 14.1).

Apart from the bone marrow and blood, other human tissues derived from constantly dividing cell populations are the gut, skin and mucous membranes found in the lungs, throat and elsewhere. So these tissues might also be suitable targets for gene therapy.[10] Many adult human tissues, though, are made up of cells that either do not divide or are dividing only very slowly. To correct gene defects in these tissues you would need to change all or at least most of the cells directly – an awesome and perhaps impossible task. It *might* be possible to treat some genetic diseases affecting non-dividing cells, however, by making blood cells produce the proteins that the non-dividing cells should be making. Even though the proteins would be made in the wrong place, they might be able to penetrate into the diseased tissue in sufficient amounts to cure, or at least lessen the severity of, some diseases.

An alternative approach to curing genetic disease might be to catch it when the sufferers are still unborn embryos made out of cells that are still actively dividing. The required genes might then be inserted into all, or at least most, of the fairly small number of embryonic cells; perhaps by infecting the embryos with suitable engineered retroviruses. Such an adventurous approach could well turn out to be feasible, but to many people it might seem much more sensible simply to abort the diseased embryos as soon as their defects are detected, and advise the parents to try again.

So we have seen various ways in which both natural viruses and viruses altered by genetic engineering are being used to carry out tasks that might otherwise be quite difficult to perform. But the exploitation of viruses is not restricted simply to finding uses for whole viruses. Possible jobs for purified viral components, especially in the delivery of drugs and other types of molecule to specific cells, are also being explored.

Jobs for membranes and proteins

One of the great problems associated with the use of many modern drugs is the difficulty of getting them into only the specific diseased cells you are trying to kill or cure. Many drugs cannot be used to their full potential, because when they become distributed throughout the entire body they damage cells other than the specific types suffering from disease. If ways could be found to *selectively* deliver drugs only to specific types of cell, then many chemicals that are currently too dangerous to use as drugs might become extremely useful; and drugs of currently limited efficacy might become much more effective.

One way to deliver a drug into a particular type of cell might be to seal the drug inside vesicles made out of the outer membrane of an appropriate enveloped virus. To achieve this, ruptured and empty viral membranes (still carrying the proteins that determine which type of cell the virus can infect) would be purified and mixed with the drug under conditions that encouraged the membranes to reseal and trap some drug molecules within them (see figure 14.2A). The drug-laden vesicles would then be administered to patients, allowing them to bind to the cells that the chosen virus could infect and thus deliver the drug into these cells alone.

One idea to make the whole process even more specific, is to incorporate into the viral membranes specific antibodies that will bind to proteins present only on the surface of diseased cells (or at least more abundant on the diseased cells). Antibodies chosen to bind to proteins found only on cancer cells, for example, might allow poisons to be delivered to cancer cells without the risk of

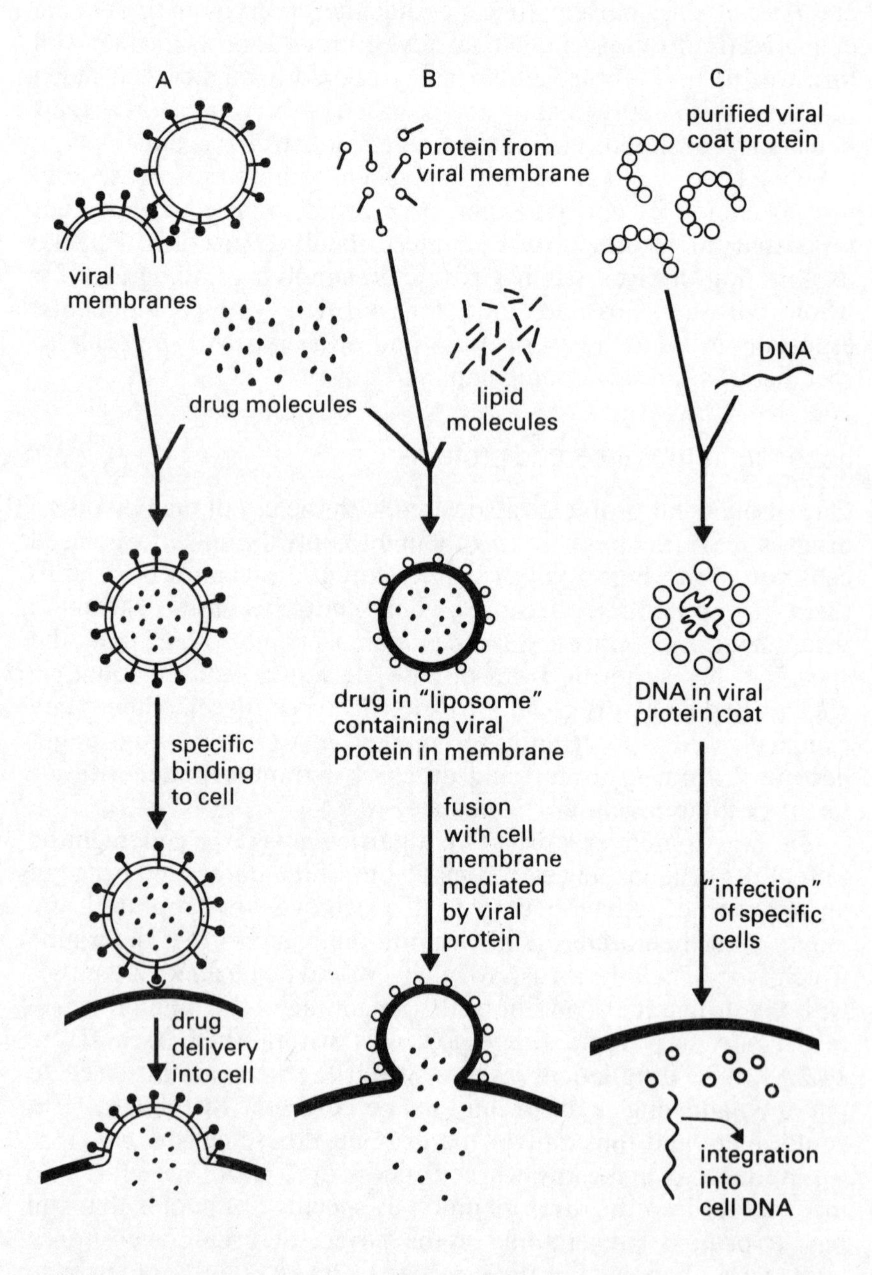

Figure 14.2 Ways to use viral membranes and proteins to deliver drugs and other chemicals into specific types of cell (see text for details)

causing any harm to a patient's healthy cells. Although there are many exciting ideas and possibilities, the development of viral membranes that serve as suitable specific drug delivery systems is still in its very early stages.[11]

In some cases it might be more appropriate not to use intact viral membranes, but instead to purify some desired proteins found in the viral membranes and then incorporate these proteins into artificial membranes. Lipid molecules can quite easily be induced to form lipid-bilayer-bound vesicles known as "liposomes"; and if these are formed in the presence of viral membrane proteins, then liposomes with *selected* properties of viral membranes may be obtained. For example, proteins that normally help viral membranes to fuse with the membranes of their host cells might be incorporated into the membranes of drug-laden liposomes to help the liposomes to deliver their drugs into cells (see figure 14.2B).[12] Once again, the addition of antibodies (or other appropriate viral proteins) to the membranes might make the delivery process much more specific.

Of course membrane systems used to deliver drugs into cells could also be used to deliver many other types of chemical into cultured cells being studied or manipulated in the laboratory. One type of chemical that scientists frequently want to deliver into cultured cells is DNA, during genetic engineering experiments designed to alter a cell's genome. Viral membranes or liposomes containing viral proteins might offer efficient ways to deliver this DNA. Another way being developed to get DNA into cultured cells exploits the outer protein coats of viruses, rather than their outer membranes.

Purified viral coat proteins, mixed with the DNA you want to transfer, can be induced to form viral coats containing the DNA (see figure 14.2C). The DNA-containing coats can then be added to suitable cells, allowing them to bind to and enter the cells and release their cargo of DNA. Once inside cells, foreign pieces of DNA can occasionally be integrated into the cell genome and expressed and replicated just like normal cellular DNA. Once again, the exploitation of purified viral protein coats to ferry new genes into cells is still in the very early stages of development,[13] but a whole range of suitable protein coats might eventually be available to transfer new genes into many different types of cell.

Weapons?

Amidst all the hopes raised by our increasing ability to control, manipulate and exploit the viruses, it must be remembered that advances in science often open up sinister possibilities in addition to their potential benefits. The advances in virology and molecular biology that will increasingly allow us to exploit the viruses in the ways discussed so far, may unfortunately also be used to cause us great harm. They might be used to convert the viruses into subtle, stealthy and devastating weapons with which to wage biological warfare.

Of course to use viruses as weapons it is not really necessary to generate weird new viruses by genetic engineering. The use of a virulent smallpox virus against an unvaccinated population, for example, might bring a nation to its knees just as effectively as any bombs or guns. But by specifically modifying viruses for use as weapons, it might be easier both to protect your own population while ensuring that the enemy is completely helpless.

The simplest use of genetic engineering to make viral weapons would involve the construction of variant forms of pre-existing dangerous viruses gathered from nature. Looking at the variation of influenza viruses has already told us that small changes in the structure of a virus's antigens can allow it to bypass any immunity raised against the original strain. By vaccinating your own troops, or even your entire population, against any variant viruses you create, it might be possible to generate biological weapons capable of killing the enemy while leaving your own people unaffected.

More adventurously, however, there has already been speculation about the possibility of constructing lethal "ethnic" viruses (or bacteria), somehow capable of selectively killing members of some particular race. It is not easy at the moment to see how such specific viruses could be reliably constructed, but we would be foolish to dismiss the possibility.

There would be little point in considering in great detail here all of the terrible things that people might be able to do to one another if our knowledge of virology were applied to such destructive ends. The important point is that the viruses certainly have the potential to be used as weapons, with either vaccination or the use of highly selective viruses being relied upon to protect the aggressor's population. One of the most worrying things about such a possibility, as with all biological warfare, is that it might allow a war to be waged unnoticed. A biological attack might be disguised to appear like a natural epidemic, at least until it had become fully established; and even if a nation realised it was the victim of biological warfare it might be difficult for it to identify its attacker.

Of course other than using viruses to attack humans, an alternative and possibly more likely approach would be to direct a viral attack against a nation's livestock or crops. If carried out with cunning such anti-agriculture warfare might allow one nation to seriously and continuously damage another nation's economy, while outwardly trading with that nation as a friend.

It would of course be a tragedy if the transformation in our relationship with the viruses were to end with us using viruses to kill and maim one another and to sabotage each others' economies, just as the age-old threat of natural infection was finally being overcome. Unfortunately, if there is one lesson that human history should teach us it is that when scientific advances have sinister potential then that potential will eventually be realised. All of the discoveries about the viruses discussed in this book, ending with our increasing ability to manipulate and exploit these simplest infectious agents of all, might therefore be the prelude to devastating biological wars, in which bacteria, fungi and other micro-organisms would probably also play a part.

We seem to have the potential either to conquer viral infection completely and turn the properties of the viruses to good use; or else to use these same properties to destroy our fellow-humans in new and ever more ingenious ways. Human wit and wisdom has made these two options available. We can only hope that same wit and wisdom allows us to make the right choice.

References

Chapter 1

1 Beijerinck, M. W., A *contagium vivum fluidum* as the cause of the mosaic disease of tobacco leaves. In *Milestones in Microbiology*, ed. T. D. Brock, Prentice-Hall, London, 1961, pp. 153–7.
2 Hughes, S. S., *The Virus: a history of the concept*. Heinemann, London, 1977.
3 Loeffler, F. and Frosch, P., Report of the commission for research on the foot-and-mouth disease. In *Milestones in Microbiology*, ed. T. D. Brock. Prentice-Hall, London, 1961, pp. 149–53.
4 Twort, F., An investigation on the nature of ultra-microscopic viruses, *The Lancet*, **II**, 1241–43, 1915.
5 d'Herelle, F., An invisible microbe that is antagonistic to the dysentery bacillus. In *Milestones in Microbiology*, ed. T. D. Brock. Prentice-Hall, London, 1961, pp. 157–9.
6 Elford, J. W., A new series of graded collodion membranes suitable for general bacteriological use, especially in filterable virus studies, *Journal of Pathology and Bacteriology*, **34**, 505–21, 1931.
7 Schlesinger, M., The Feulgen reaction of the bacteriophage substance, *Nature*, **138**, 508–9, 1936.
8 Bawden, F. C. and Pirie, N. W., The isolation and some properties of liquid crystalline substances from Solanaceous plants infected with three strains of tobacco mosaic virus, *Proceedings of the Royal Society of London, Series B*, **123**, 274–320, 1937.

Chapter 2

1 Further details of the basic biochemistry covered in this chapter can be found in textbooks such as Roberts, M. B. V., *Biology*, Thomas Nelson, Walton-on Thames, 1982; and Lehninger, A. L., *Biochemistry*, Worth, New York, 1975.

Chapter 3

1 Stanley, W. M., Isolation of a crystalline protein possessing the properties of tobacco mosaic virus, *Science*, **81**, 644–5, 1935.

2 Watson, J. D. and Crick, F. H. C., Molecular structure of nucleic acids – a structure for deoxyribose nucleic acid, *Nature*, **171**, 737–8, 1953.

3 Madeley, C. R., *Virus Morphology*, Churchill Livingstone, Edinburgh and London, 1972.

4 Klug, A., Architectural design of spherical viruses, *Nature*, **303**, 378–9, 1983.

5 Gierer, A. and Schramm, G., Infectivity of ribonucleic acid from tobacco mosaic virus, *Nature*, **177**, 702–3, 1956.

6 Fraenkel-Conrat, H., The role of the nucleic acid in the reconstitution of active tobacco mosaic virus, *Journal of the American Chemical Society*, **78**, 882–3, 1956.

Chapter 4

1 Dimmock, N. J., Initial stages in infection with animal viruses, *Journal of General Virology*, **59**, 1–22, 1982.

2 Bukrinskaya, A. G., Penetration of viral genetic material into host cell, *Advances in Virus Research*, **27**, 141–204, 1982.

3 Matlin, K. S., Reggio, H., Helenius, A. and Simons, K., Infectious entry pathway of influenza virus in a canine kidney cell line, *Journal of Cell Biology*, **91**, 601–13, 1981.

4 Simons, K., Garoff, H. and Helenius, A., How an animal virus gets into and out of its host cell, *Scientific American*, February 1982, pp. 46–54.

5 Marx, J. L., First parvovirus linked to human disease, *Science*, **223**, 152–3, 1984.

6 Simpson, R. W., McGinty, L., Simon, L., Smith, C. A., Godzeski, C. W. and Boyd, R. J., Association of parvoviruses with rheumatoid arthritis of humans, *Science*, **223**, 1425–8, 1984.

7 Astell, C. R., Thomson, M., Merchlinsky, M. and Ward, D. C., The complete DNA sequence of a minute virus of mice, an autonomous parvovirus, *Nucleic Acids Research*, **11**, 999–1018, 1983.

8 Pintel, D., Dadachanji, D., Astell, C. R. and Ward, D. C., The genome of minute virus of mice encodes two overlapping transcription units, *Nucleic Acids Research*, **11**, 1019–38, 1983.

9 McCauley, J. W. and Mahy, B. W. J., Structure and function of the influenza virus genome, *Biochemical Journal*, **211**, 281–94, 1983.

10 Roizman, B., The organization of the herpes simplex virus genomes, *Annual Review of Genetics*, **13**, 25–57, 1979.

11 Wittek, R., Organization and expression of the poxvirus genome, *Experentia*, **38**, 285–410, 1982.

12 Fraenkel-Conrat, H. and Williams, R. C., Reconstitution of active tobacco mosaic virus from its inactive protein and nucleic acid components, *Proceedings of the National Academy of Sciences of the U.S.A.*, **41**, 690–8, 1955.

Chapter 5

1 Lwoff, A., The prophage and I. In *Phage and the Origins of Molecular Biology*, eds J. Cairns, G. S. Stent and J. D. Watson. Cold Spring Harbor Laboratory, New York, 1966, pp. 88–99.
2 Herskowitz, I. and Hagen, D., The lysis-lysogeny decision of phage lambda: explicit programming and responsiveness, *Annual Review of Genetics*, **14**, 399–445, 1980.
3 Nash, H. A., Integration and excision of bacteriophage lambda: the mechanism of conservative site specific recombination, *Annual Review of Genetics*, **15**, 143–67, 1981.
4 Weinberg, R. A., Integrated genomes of animal viruses, *Annual Review of Biochemistry*, **49**, 197–226, 1980.
5 Sambrook, J., Westphal, H., Srinivasan, P. R. and Dulbecco, R., The integrated state of viral DNA in SV40-transformed cells, *Proceedings of the National Academy of Sciences of the U.S.A.*, **60**, 1288–95, 1968.
6 Varmus, H. E., Form and function of retroviral proviruses, *Science*, **216**, 812–20, 1982.
7 Hughes, S. H., Synthesis, integration and transcription of the retroviral provirus, *Current Topics in Microbiology and Immunology*, **103**, 23–49, 1983.
8 Jaenisch, R., Endogenous retroviruses, *Cell*, **32**, 5–6, 1983.
9 O'Brien, S. J., Bonner, T. I., Cohen, M., O'Connell, C. and Nash, W. G., Mapping of an endogenous retroviral sequence to human chromosome 18, *Nature*, **303**, 74–7, 1983.

Chapter 6

1 Mims, C. A., Entry of micro-organisms into the body. In Mims, C. A., *The Pathogenesis of Infectious Disease*, Academic Press, London and New York, 1982, pp. 8–43.
2 Fenner, F. and White, D. O., *Medical Virology*, Academic Press, New York and London, 1976, pp. 182–8.
3 Mims, C. A., The spread of microbes through the body. In Mims, C. A., *The Pathogenesis of Infectious Disease*, Academic Press, London and New York, 1982, pp. 82–108.

Chapter 7

1 Mims, C. A., The encounter of the microbe with the phagocytic cell.

In Mims, C. A., *The Pathogenesis of Infectious Disease*, Academic Press, London and New York, 1982, pp. 56–81.
2 Herberman, R. B. and Ortaldo, J. R., Natural killer cells: their role in defenses against disease, *Science*, **214**, 24–30, 1981.
3 Gordon, J. and Minks, M. A., The interferon renaissance: molecular aspects of induction and action, *Microbiological Reviews*, **45**, 244–66, 1981.
4 Whitaker-Dowling, P. A., Wilcox, D. K., Widnell, C. C. and Youngner, J. S., Interferon-mediated inhibition of virus penetration, *Proceedings of the National Academy of Sciences of the U.S.A.*, **80**, 1083–6, 1983.
5 Strannegard, O., Mechanisms of defence against virus infections, in *Textbook of Medical Virology*, eds. E. Lycke and E. Norrby, Butterworths, London, 1983, pp. 178–88.
6 Onions, D. E., The immune response to virus infections, *Veterinary Immunology and Immunopathology*, **4**, 237–77, 1983.

Chapter 8

1 Fenner, F., Biological control, as exemplified by smallpox eradication and myxomatosis, *Proceedings of the Royal Society of London, Series B*, **218**, 259–85, 1983.
2 Lycke, E. and Norrby, E., Virus-induced changes of cell structures and functions. In *Textbook of Medical Virology*, eds. E. Lycke and E. Norrby, Butterworths, London, 1983, pp. 93–104.
3 Luria, S. E., Darnell, J. E., Baltimore, D. and Campbell, A., *General Virology*, John Wiley and Sons, New York, 1978, pp. 391–433.
4 Oldstone, B. A., Rodriguez, M., Daughaday, W. H. and Lampert, P. W., Viral perturbation of endocrine function: disordered cell function leads to disturbed homeostasis and disease, *Nature*, **307**, 278–81, 1984.
5 Fields, B. N., Viruses and tissue injury, *Nature*, **307**, 213–14, 1984.
6 Mims, C. A., *The Pathogenesis of Infectious Disease*, Academic Press, London and New York, 1982, pp. 191–2.
7 Narayan, O., Herzog, S., Frese, K., Scheefers, H. and Rott, R., Behavioural disease in rats caused by immunopathological responses to persistent Borna virus in the brain, *Science*, **220**, 1401–3, 1983.
8 Shusterman, N. and London. W. T., Hepatitis B and immune-complex disease, *New England Journal of Medicine*, **310**, 43–5, 1984.
9 Fujinami, R. S., Oldstone, B. A., Wroblewska, Z., Frankel, M. E. and Koprowski, H., Molecular mimicry in virus infection: crossreaction of measles virus phosphoprotein or of herpes simplex virus protein with human intermediate filaments, *Proceedings of the National Academy of Sciences of the U.S.A.*, **80**, 2346–50, 1983.

Chapter 9

1 Hermodsson, S., Orthomyxoviruses (influenza viruses). In *Textbook of Medical Virology*, eds. E. Lycke and E. Norrby. Butterworths, London, 1983, pp. 262–71.
2 Van Rompuy, L., Min Jou, W., Verhoeyen, M., Huylebroeck, D. and Fiers, W., Molecular variation of influenza surface antigens, *Trends in Biochemical Sciences*, **8**, 414–17, 1983.
3 Stroop, W. G. and Baringer, J. R., Persistent, slow and latent viral infections, *Progress in Medical Virology*, **28**, 1–43, 1982.
4 Lycke, E., Herpes viruses. In *Textbook of Medical Virology*, eds. E. Lycke and E. Norrby. Butterworths, London, 1983, pp. 304–22.
5 Johnson, R. T., Subacute sclerosing panencephalitis. In *Cecil Textbook of Medicine*, eds. J. B. Wyngaarden and L. H. Smith, W. B. Saunders, Philadelphia, 1982, pp. 2101–2.
6 Fenner, F. J. and White, D. O., *Medical Virology*, Academic Press, New York and London, 1976, pp. 439–48.
7 Hoofnagle, J. H., Type B hepatitis: virology, serology and clinical course, *Seminars in Liver Disease*, **1** 7–14, 1981.

Chapter 10

1 Ellerman, V. and Bang, O., Experimentelle leukamie bei huhern, *Zentralblatt fur Bakteriologie*, *Parasitenkunde*, *Infektionskrank-heiten und Hygiene*, **46**, 595–609, 1908.
2 Rous, P., A sarcoma of the fowl transmissible by an agent separable from tumor cells, *Journal of Experimental Medicine*, **13**, 397–411, 1911.
3 Gye, W. E. and Purdy, W. J., *The Cause of Cancer*, Cassell, London, 1931.
4 Bittner, J. J., The genesis of breast cancer in mice, *Texas Reports on Biology and Medicine*, **10**, 160–6, 1952.
5 Axel, R., Schlom, J. and Spiegelman, S., Presence in human breast cancer of RNA homologous to mouse mammary tumour virus RNA, *Nature*, **235**, 32–6, 1972.
6 Moore, D. H., Moore II, D. H. and Moore, C. T., Breast carcinoma etiological factors, *Advances in Cancer Research*, **40**, 189–253, 1983, pp. 223–7.
7 Gardner, M. B., Viruses as environmental carcinogens: an agricultural perspective, *Basic Life Sciences*, **21**, 171–87, 1982.
8 Waterson, A. P., Human cancers and human viruses, *British Medical Journal*, **284**, 446–8, 1982.
9 Szmuness, W., Hepatocellular carcinoma and the hepatitis B virus: evidence for a causal association, *Progress in Medical Virology*, **24**, 40–69, 1978.
10 Beasley, R. P., Lin, C. C., Hwang, L. Y. and Chien, C. S.,

Hepatocellular carcinoma and hepatitis B virus, *The Lancet*, **II**, 1129–32, 1981.
11 Zuckerman, A. J., Primary hepatocellular carcinoma and hepatitis B virus, *Transactions of the Royal Society of Tropical Medicine and Hygiene*, **76**, 711–18, 1982.
12 Lutwick, L. I. and Robinson, W. S., DNA synthesized in the hepatitis B Dane particle DNA polymerase reaction, *Journal of Virology*, **21**, 96–104, 1977.
13 Shafritz, D. A., Shouval, D., Sherman, H. I., Hadziyannis, S. J. and Kew., Integration of hepatitis B virus DNA into the genome of liver cells in chronic liver disease and hepatocellular carcinoma, *New England Journal of Medicine*, **305**, 1067–73, 1981.
14 Prevention of liver cancer, *WHO Technical Reports*, No. 691, 9, 1983.
15 Wyke, J. A., Oncogenic viruses, *Journal of Pathology*, **135**, 39–85, 1981, (pp. 51–5).
16 Wong-Staal, F., Hahn, B., Manzari, V., Colombini, S., Franchini, G., Gelmann, E. P. and Gallo, R. C., A survey of human leukaemias for sequences of a human retrovirus, *Nature*, **302**, 626–8, 1983.
17 Dermer, G. B., Human cancer research, *Science*, **221**, 318, 1983.
18 Croce, C. M., Integration of oncogenic viruses in mammalian cells, *International Review of Cytology*, **71**, 1–17, 1981.
19 Bishop, J. M., Oncogenes, *Scientific American*, March 1982, pp. 69–78.
20 Doolittle, R. F., Hunkapiller, M. W., Hood, L. E., Devare, S. G., Robbins, K. C., Aaronson, S. A. and Antionades, H. N., Simian sarcoma virus *onc* gene, v-*sis*, is derived from the gene (or genes) encoding a platelet-derived growth factor, *Science*, **222**, 275–7, 1983.
21 Waterfield, M. D., Scarce, G. T., Whittle, N., Stroobant, P., Johnsson, A., Wasteson, A., Westermark, B., Heldin, C. H., Huang, J. S., and Deuel, T. F., Platelet-derived growth factor is structurally related to the putative transforming protein $p28^{sis}$ of simian sarcoma virus, *Nature*, **304**, 35–9, 1983.
22 Deuel, T. F., Huang, J. S., Huang, S. S., Stroobant, P. and Waterfield, M. D., Expression of a platelet-derived growth factor-like protein in simian sarcoma virus transformed cells, *Science*, **221**, 1348–50, 1983.
23 Downward, J., Yarden, Y., Mayes, E., Scrace, G., Totty, N., Stockwell, P., Ullrich, A., Schlessinger, J. and Waterfield, M. D., Close similarity of epidermal growth factor receptor and v-*erb*-B oncogene protein sequences, *Nature*, **307**, 521–7, 1984.
24 Bishop, J. M., Cellular oncogenes and retroviruses, *Annual Review of Biochemistry*, **52**, 301–54, 1983.
25 Steffen, D., Proviruses are adjacent to c-*myc* in some murine leukemia virus-induced lymphomas, *Proceedings of the National Academy of Sciences of the U.S.A.*, **81**, 2097–101, 1984.
26 Hardy, W. D., A new package for an old oncogene, *Nature*, **308**, 775, 1984.

27 Land, H., Parada, L. F. and Weinberg, R. A., Cellular oncogenes and multistep carcinogenesis, *Science*, **222**, 771–8, 1983.
28 Murphree, A. L. and Benedict, W. F., Retinoblastoma: clues to human oncogenesis, *Science*, **223**, 1028-33, 1984.

Chapter 11

1 Jenner, E., An inquiry into the causes and effects of the variolae vaccinae, a disease discovered in some of the western counties of England, particularly Gloucestershire, and known by the name of the cow pox. In *Milestones in Microbiology*, ed. T. D. Brock. Prentice-Hall, London, 1961, pp. 121–5.
2 Behbehani, A. M., The smallpox story: life and death of an old disease, *Microbiological Reviews*, **47**, 455–509, 1983.
3 Fenner, F., Biological control, as exemplified by smallpox eradication and myxomatosis, *Proceedings of the Royal Society of London, Series B*, **218**, 259–85, 1983.
4 Cherfas, J., *Man Made Life*, Basil Blackwell, Oxford, 1982.
5 McAleer, W. J., Buynak, E. B., Maigetter, R. Z., Wampler, D. E., Miller, W. J. and Hilleman, M. R., Human hepatitis B vaccine from recombinant yeast, *Nature*, **307**, 178–80, 1984.
6 Linnebank, G., Biogen's inside track, *Nature*, **304**, 297, 1983.
7 Crosnier, J., Jungers, P., Courouce, A. M., Laplanche, A., Benhamou, E., Degos, F., Lacour, B., Prunet, P., Cerisier, Y. and Guesry, P., Randomised placebo-controlled trial of hepatitis B surface antigen vaccine in French haemodialysis units: 1, medical staff, *The Lancet*, **I**, 455–9, 1981.
8 Lerner, R. A., Synthetic vaccines, *Scientific American*, February 1983, pp. 48–56.
9 Newmark, P., Will peptides make vaccines?, *Nature*, **305**, 9, 1983.
10 Williams, N., Building new vaccines, *Nature*, **306**, 427, 1983.
11 Roizman, B., Warren, J., Thuning, C. A., Fanshaw, M. S., Norrild, B. and Meignier, B., Application of molecular genetics to the design of live herpes simplex virus vaccines, *Developments in Biological Standardization*, **52**, 287–304, 1982.

Chapter 12

1 Liu, C., Antiviral drugs, *Medical Clinics of North America*, **66**, 235–44, 1982, p. 235.
2 Elion, G. B., The biochemistry and mechanism of action of acyclovir, *Journal of Antimicrobial Chemotherapy*, **12**, Supplement B: 9–17, 1983.
3 *Journal of Antimicrobial Chemotherapy*, **12**, Supplement B: whole issue (entitled "Acyclovir"), 1983.

4 Cheng, Y. C., Huang, E. S., Lin, J. C., Mar, E. C., Pagano, J. S., Dutschman, G. E. and Grill, S. P., Unique spectrum of activity of 9–[(1,3-dihydroxy-2-propoxy)methyl]-guanine against herpesviruses *in vitro* and its mode of action against herpes simplex virus type 1, *Proceedings of the National Academy of Sciences of the U.S.A.*, **80**, 2767–70, 1983.
5 Lin, J. C., Smith, M. C., Cheng, Y. C. and Pagano, J. S., Epstein-Barr virus: inhibition of replication by three new drugs, *Science*, **221**, 578–9, 1983.
6 Oxford, J. S., and Galbraith, A., Antiviral activity of amantadine: a review of laboratory and clinical data, *Pharmacology and Therapeutics*, **11**, 181–262, 1980.
7 Ishitsuka, H., Ninomiya, T., Ohsawa, C., Ohiwa, T., Fujiu, M., Umeda, I., Shirai, H. and Suhara, Y., New antirhinovirus agents, Ro 09–0410 and Ro 09–0415, in *Current Chemotherapy and Immunotherapy*, vol. 2, eds. P. Periti and G. Gialdroni-Grassi. American Society for Microbiology, Washington, 1982, pp. 1083–5.
8 The Medical Research Council's Scientific Committee on Interferon, Effect of interferon on vaccination in volunteers, *The Lancet*, **I**, 873–5, 1962.
9 Merigan, T. C., Human interferon as a therapeutic agent – current status, *New England Journal of Medicine*, **308**, 1530–1, 1983.
10 (Anon.), "Toxicity" of interferon, *The Lancet*, **I**, 1256, 1983.
11 (Anon.), Further doubts cast over interferon's future, *Chemistry and Industry*, 3 October 1983, p. 725.
12 Sharma, O. K., Engels, J., Jager, A., Crea, R., van Broom, J. and Goswami, B. B., 3′-O-methylated analogs of 2–5A as inhibitors of virus replication, *FEBS Letters*, **158**, 298–300, 1983.
13 Chustecka, Z., New drug for genital herpes, *New Scientist*, 12 April 1984, p. 22.
14 De Clercq, E., Specific targets for antiviral drugs, *Biochemical Journal*, **205**, 1–13, 1982.
15 Richardson, C. D. Scheid, A. and Choppin, P. W., Specific inhibition of paramyxovirus and myxovirus replication by oligopeptides with amino acid sequences similar to those at the N-termini of the F_1 or HA_2 viral polypeptides, *Virology*, **105**, 205–22, 1980.
16 Zamecnik, P. C. and Stephenson, M. L., Inhibition of Rous sarcoma virus replication and cell transformation by a specific oligodeoxynucleotide, *Proceedings of the National Academy of Sciences of the U.S.A.*, **75**, 280–4, 1978.
17 Skinner, G. R. B., Lithium ointment for genital herpes, *The Lancet*, **II**, 288, 1983.
18 Eby, G. A., Davis, D. R. and Halcomb, W. W., Reduction in duration of common colds by zinc gluconate lozenges in a double-blind study, *Antimicrobial Agents and Chemotherapy*, **25**, 20–4, 1984.

Chapter 13

1 Hershey, A. D. (ed.), *The Bacteriophage Lambda*, The Cold Spring Harbor Laboratory, New York, 1971.
2 Pinching, A. J., Acquired immune deficiency syndrome, *Hospital Update*, **10**, 117–29, 1984.
3 Marx, J. L., Strong new candidate for AIDS agent, *Science*, **224**, 475–7, 1984.
4 Joyce, C., How to safeguard against AIDS, *New Scientist*, 31 May 1984, p. 8.
5 Lifson, J. D., Benike, C. J., Mark, D. F., Koths, K. and Engleman, E. G., Human recombinant interleukin-2 partly reconstitutes deficient in vitro immune responses of lymphocytes from patients with AIDS, *The Lancet*, **I**, 698–702, 1984.
6 Yoon, J. W., Austin, M., Onodera, T. and Notkins, A. L., Virus-induced diabetes mellitus, *New England Journal of Medicine*, **300**, 1173–9, 1979.
7 Simpson, R. W., McGinty, L., Simon, L., Smith, C. A., Godzeski, C. W. and Boyd, R. J., Association of parvoviruses with rheumatoid arthritis of humans, *Science*, **223**, 1425–8, 1984.
8 Billings, P. B., Hoch, S. O., White, P. J., Carson, D. A. and Vaughan, J. H., Antibodies to Epstein–Barr virus nuclear antigen and to rheumatoid arthritis nuclear antigen identify the same polypeptide, *Proceedings of the National Academy of Sciences of the U.S.A.*, **80**, 7104–8, 1983.
9 (Anon.), The viral aetiology of rheumatoid arthritis, *The Lancet*, **I**, 772–3, 1984.
10 Benditt, E. P., Barett, T. and McDougall, J. K., Viruses in the etiology of atherosclerosis, *Proceedings of the National Academy of Sciences of the U.S.A.*, **80**, 6386–9, 1983.
11 Behan, P. O., Postviral neurological syndromes, *British Medical Journal*, **287**, 853–4, 1983.
12 Mims, C., Multiple sclerosis — the case against viruses, *New Scientist*, 30 June 1983, pp. 938–40.
13 Kimberlin, R. H., Scrapie agent: prions or virinos?, *Nature*, **297**, 107–8, 1982.
14 Rohwer, R. G., Scrapie infectious agent is virus-like in size and susceptibility to inactivation, *Nature*, **308**, 658–62, 1984.
15 Diener, T. O., Viroids, *Scientific American*, January 1981, pp. 58–65.
16 Coggin, J. H., Bellomy, B. B., Tjomas, K. V. and Pollock, W. J., B-cell and T-cell lymphomas and other associated diseases induced by an infectious DNA viroid-like agent in hamsters (*Mesocricetus auratus*), *American Journal of Pathology*, **110**, 254–66, 1983.
17 Matthews, R. E. F., The origin of viruses from cells, *International Review of Cytology*, Supplement No. 15; pp. 245–280, 1983.
18 Diener, T. O., Viroids and their interactions with host cells, *Annual Review of Microbiology*, **36**, 239–58, 1982, (p. 254).

19 Zimmern, D., Do viroids and RNA viruses derive from a system that exchanges genetic information between eukaryotic cells?, *Trends in Biochemical Sciences*, **7**, 205–7, 1982.
20 Kiefer, M. C., Owens, R. A. and Diener, T. O., Structural similarities between viroids and transposable genetic elements, *Proceedings of the National Academy of the U.S.A.*, **80**, 6234–8, 1983.

Chapter 14

1 Slopek, S., Durlakowa, I., Weber-Dabrowska, B., Kucharewicz-Krukowsca, A., Dabrowski, M. and Bisikiewicz, R., Results of bacteriophage treatment of suppurative bacterial infections, *Archivum Immunologiae et Therapie Experimentalis*, **31**, 267–327, 1983.
2 Williams Smith, H. and Huggins,M. B., Effectiveness of phages in treating experimental *Escherichia coli* diarrhoea in calves, piglets and lambs, *Journal of General Microbiology*, **128**, 2659–75, 1983.
3 Winger, L., Winger, C., Shastry, P., Russell A. and Longenecker, M., Effective generation *in vitro*, from human peripheral blood cells, of monoclonal Epstein–Barr virus transformants producing specific antibody to a variety of antigens without prior deliberate immunization, *Proceedings of the National Academy of Sciences of the U.S.A.*, **80**, 4484–8, 1983.
4 Smith, G. L., Godson, G. N., Nussenzweig, V., Nussenzweig, R. S., Barnwell, J. and Moss, B., *Plasmodium knowlesi* sporozoite antigen: expression by infectious recombinant vaccinia virus, *Science*, **224**, 397–9, 1984.
5 Cherfas, J., *Man Made Life*, Basil Blackwell, Oxford, 1982, pp. 90–5.
6 Kolata, G., Gene therapy method shows promise, *Science*, **223**, 1376–9, 1984.
7 Marx, J. L., Specific expression of transferred genes, *Science*, **222**, 1001–2, 1983.
8 Howard, B. H., Vectors for introducing genes into cells of higher eukaryotes, *Trends in Biochemical Sciences*, **8**, 209–12, 1983.
9 Anderson, W. F., and Diacumakos, E. G., Genetic engineering in mammalian cells, *Scientific American*, July 1981, pp. 60–93.
10 Cline, M. J., Genetic engineering of mammalian cells: its potential application to genetic diseases of man, *Journal of Laboratory and Clinical Medicine*, **99**, 299–308, 1982.
11 Sargiacomo, M., Barbieri, L., Stirpe, F. and Tomasi, M., Cytotoxicity acquired by ribosome-inactivating proteins carried by reconstituted Sendai virus envelopes. *FEBS Letters*, **157**, 150–4, 1983.
12 White, J., Helenius, A. and Gething, M. J., Haemagglutinin of influenza virus expressed from a cloned gene promotes membrane fusion, *Nature*, **300**, 658–9, 1982, (p. 659).
13 Slilaty, S. N. and Aposhian, H. V., Gene transfer by polyoma-like particles assembled in a cell-free system, *Science*, **220**, 725–7, 1983.

Further Reading

There are no good introductory books about the viruses at a similar level to this one (that's why I wrote it!) but anyone wishing to delve further into the subjects I have introduced will find more detailed information in the following:

An Introduction to the History of the Viruses, by A. P. Waterson and L. Wilkinson, Cambridge University Press, 1978 – a readable account of the earliest work in virology written by two experts.

Introduction to Modern Virology, by S. B. Primrose and N. J. Dimmock, Blackwell Scientific, 1980 – a detailed technical introduction to the molecular biology of viruses and virus infection.

General Virology, by S. E. Luria, J. E. Darnell Jr., D. Baltimore and A. Campbell, John Wiley & Sons, 1978 – a good comprehensive textbook of virology, packed with information but certainly not light reading!

Medical Virology, by F. Fenner and D. O. White, Academic Press, 1976 – a textbook aimed at medical students, filled with useful information although in need of updating as a new edition.

Textbook of Medical Virology, edited by E. Lycke and E. Norrby, Butterworths, 1983 – a collection of detailed essays aimed at the specialist. Interesting and up-to-date.

Viral Pathogenesis and Immunology, by C. A. Mims and D. O. White, Blackwell Scientific, 1984 – a comprehensive survey of how viruses cause disease and how the body fights them off, aimed at the specialist.

Index

acyclovir, 176–9
adenine, 16, 17, 26
adenosine deaminase, 220, 221
adenoviruses, 37, 42
Agastache folium, 180
AIDS, 189–95
  virus, 193–4
alimentary tract, 76, 79, 80, 81,84, 96, 112, 224
alphaviruses, 42
amantadine, 179–80
amino acids, 20–8, 54, 170, 171, 215
antibodies, *see* immune system, antibodies of
antigens, 92, 93, 94, 95, 96, 106, 168–74
  of influenza viruses, 116–19
  see also immune system; vaccines
arenaviruses, 42
arthritis, rheumatoid, 55, 197–8
aspirin, 114
assembly of viruses, 57–8
atherosclerosis, 198–9

Baccillus megaterium, 62
bacteria, 111, 116, 213, 214
  attack of by viruses, 6–7, 39–40, 61–7, 218–19; *see also* bacteriophages
  conjugation of, 64
  in genetic engineering, 87, 168, 220
  size of, 9
bacteriophages,
  as antibiotics, 6–7, 218–19
  assembly of, 58
  classification of, 41
  cultivation of, 8–10
  discovery of, 6–7
  in genetic engineering, 220
  infection by, 39–40
  lambda, 64–7, 188
  lysogeny and, 61–7
Bang, Olaf, 132
barrier systems, 81–1, 85, 90
bases, 15–18, 27
  see also specific bases
Beasley, R. Palmer, 137
Beijerinck, Martinus, 2–5, 8
Benditt, Earl, 198
Biogen, 168
biological warfare, *see* weapons
bites, 78, 79
Bittner, John, 133, 134
blindness, 76, 79
blood, 76, 78, 80, 82, 84, 85, 130, 191, 192
  cells, 85, 86–96, 114, 120
  in gene therapy, 221–5
  vessels, 82, 83, 105, 115, 198–9
bone marrow, 85, 221, 222, 224
Borna virus, 105
brain, 81, 84, 113, 114, 115, 116, 122
  diseases of, 116, 127–8, 199–204
bronchiolitis, 42
Burkitt's lymphoma, 139, 140
Burroughs Wellcome Company, 176

cancer, 120, 127, 132–58, 174, 182, 190, 206
  of blood, 139
  of breast, 133–4
  cells, 141–5, 225
  of cervix, 140
  genes, 146–58
  of genitals, 140
  and interferon, 87, 89
  of liver, 136–9, 140
  proteins, 146–58
  of skin, 140
  suppressor genes, 158
virally induced, 65, 69, 71, 102, 110, 132–58
carbohydrates, 13, 31–2, 185
  of viruses, 10, 12, 35, 54, 60
  see also glycoproteins
carcinogens, 71, 138, 142, 155, 156
cells,
  basic biochemistry of, 12–32
  culture of, 9
  damage of by viruses, 99–108
  destruction of by immune system, 59, 86, 87, 92, 103–8, 128, 197, 200
  infection of, 44–8, 79–81; *see also* infection
  lysis of, 6, 7, 10, 59, 62, 63, 64
centrifugation, 10, 68
chicken pox, 42, 77, 124, 126–7
Choppin, Purnell, 184
chromosomes,
  bacterial, 64, 66, 67, 68
  cell, 70, 71, 101, 103
classification of viruses, 41–2
codons, *see* genetic code
cold, common, 37, 42, 75, 76, 77, 79, 120
  drugs against, 180, 182, 187
cold sores, 42, 125
conjunctiva, 76, 79
complement, 93, 94, 95, 105
coronaviruses, 42
coughs, 113
cowpox, 42, 160
coxsackieviruses, 42, 196
Creutzfeld–Jakob disease, *see* Jakob–Creutzfeld disease
Crick, Francis, 19
croup, 42
cultivation of viruses, 8–10
cytomegalovirus, 42, 140
cytosine, 16, 17

defective viruses, 121, 122, 128, 150, 153, 223
delirium, *see* fever
dengue, 42
deoxyadenosine phosphate, 15
deoxycytidine phosphate, 15
2–deoxyglucose, 186
deoxyguanosine phosphate, 15
deoxyribonucleic acid, *see* DNA
diabetes, 196–7
discovery of viruses, 1–11
disease, 99, 100, 109–31
  see also specific diseases
DNA, 12–20
  infectious, 206, 208–13
  see also genes; genetic code; integration of DNA; plasmids; replication; transcription; transposons
DNA polymerase, 177, 178
Doolittle, Russell, 147
drug addicts, 78, 191
drugs,
  anti-viral, 57, 175–87
  delivery by viral components, 225–7
  see also specific drugs

echoviruses, 42
Edwin Burgess Limited, 184
electron microscopy, 34–5, 45, 47, 59
Elford, William, 7–8
Ellerman, Vilhelm, 132
embryos,
  gene therapy of, 224–5
  infection of, 128–9
encephalitis, 42, 127
endocytosis, *see* entry of viruses into cells
endogenous pyrogen, 114; *see also* fever
endosomes, 46, 47
entry of viruses into cells, 39–40, 44–8
enzymes, 13, 20–4, 26, 100, 101, 102
  attack of by drugs, 176–9, 185, 186

of cells, 48, 54, 55, 56
viral, 48, 51, 54, 55–7, 66, 69, 70
see also complement
Epstein–Barr virus, 42, 139, 140, 198, 219
see also glandular fever
erythema infectiosum, 55
Escherichia coli, 64, 66, 67
evolution, 13, 23, 45, 72, 73, 86, 106
of fever, 115
of influenza viruses, 117–19
of viruses, 97, 98, 99, 100
see also origin of viruses

faeces, 78
fat, *see* lipids
fever, 113–15, 125, 126
flaviviruses, 42
food, 78
foot-and-mouth disease virus, 5, 7–9, 172
Frosch, Paul, 5

Gallo, Robert, 193
Gardner, Murray, 140
gastro-enteritis, 42, 111
gene therapy, 220–5
genes, 25–8
recombination of, 64
split, 209, *see also* introns
viral, 48, 49, 55–7, 207–16
see also cancer genes; gene therapy; genetic code; genetic engineering; integration of DNA
genetic code, 19, 20, 24–8, 215
genetic diseases, *see* gene therapy
genetic engineering, 87, 168–70, 172–4, 182, 227
in warfare, 228–9
use of viruses in, 220–5, 227, 228–9
genome, 25
German measles, *see* rubella
germ cells, infection of, 71
germ theory of disease, 1
glandular fever, 42, 78
see also Epstein–Barr virus
glucosamine, 186
glycoproteins, 31, 38, 44, 56, 60, 92, 185
of influenza viruses, 117–19
glycosylation inhibitors, 185, 186
growth factors, 147, 148, 149
guanine, 16, 17
Gye, William, 133

haemagglutinin, of influenza viruses, 117–19
haemoglobin, 220, 221
headache, 113, 115
heart, 116
see also atherosclerosis
helical viral architecture, 35
hepatitis A virus, 42
hepatitis B, 42, 78, 104, 105, 129–31, 192, 193
and cancer, 136–9
treatment of, 181, 182
vaccination against, 136, 138, 139, 167, 168–70, 172, 173
hepatitis B virus, 42, 78, 130, 136–9, 192, 193
see also hepatitis B
Herelle, Felix d', 6
"herpes", 42, 79, 123–6, 182
see also herpesviruses
herpesviruses, 56–7, 103, 104, 106, 167, 173
and blindness, 76, 79
and cancer, 139, 140
drugs against, 176–9, 182, 184, 187
simplex, 42, 79, 125, 126, 127, 140, 198–9
size, 9
structure, 42
see also "herpes"
hormones, 102, 134, 138
HTLV, 139, 157, 193, 194
hypothalamus, 114

icosahedral viral architecture, 36–7
immune system, 85–96, 100, 103–8, 120, 121, 122, 123, 125, 129
antibodies of, 81, 82, 90, 93–4, 95, 96, 106, 121, 122, 128, 129, 130, 181, 219, 225–7
autoimmunity and, 92, 105–7, 128, 167, 197, 200
B-cells of, 93–4, 95, 96, 106, 219

deficiency of, 120, 121, 122, 123, 126, 127, 128, 130, 221, *see also* AIDS
immune complexes of, 105
memory cells of, 95–6
non-specific, 79, 80, 81, 86–9
specific, 90–6
T-cells of, 90–3, 95, 96, 106, *see also* AIDS
see also antigens; complement; vaccines
immunisation, 90, 95–6, 99, 113
see also immune system
imunovir, 183, 194
infection, 75–84
acute, 109, 110–16, 119, 120, 125, 126, 127, 129, 130
inapparent, 98, 99, 109, 110, 130
persistent, 69, 105, 109, 110, 120–31, 197, 199
spread of, 81–4
tissue specificity of, 79–81
see also entry of viruses into cells
inflammation, 91, 92, 95, 104–6, 125
influenza, 42, 110–19
see also influenza viruses
influenza viruses, 55–6
drugs against, 179–80
entry into cells, 46, 47–8
structure of, 37–8, 42, 81
transmission of, 77
vaccines against, 163, 166, 167, 172, 173
variation of, 116–19
see also influenza
insulin, 196–7
integration of DNA, 54, 61–74, 123, 227
animal DNA virus, 65, 68–9
bacteriophage, 61–7
and cancer, 145–6, 153–4
retroviral, 69–71
interferon, 87–9, 90, 91, 92, 168
as a drug, 181–3, 194
interleukin, 194
intestine, *see* alimentary tract
introns, 208, 209, 210, 211
Ivanovski, Dimitri, 1–5

Jakob-Creutzfeld disease, 200–4
jaundice, 130
see also hepatitis A virus; hepatitis B virus
Jenner, Edward, 159–61

Kaposi's sarcoma, 140, 190
kidneys, 84, 105
Koch, Robert, 1, 135

lassa fever, 42
latency, 122, 123, 125, 126, 127
LCM virus, 104
Leeuwenhoek, Antony van, 1
Lesch-Nyhan syndrome, 221
life, definition of, 72–4
lipids, 13, 28–31, 227
as viral components, 10, 12, 54
see also membranes
liposomes, 227
lithium, as an anti-viral, 186–7
Liu, Chien, 175
liver, 84, 104, 130, 136–9
Loeffler, Friedrich, 5
Lutwick, Larry, 137
Lwoff, André, 61–4
lymph, 5, 82, 83, 86, 92
lymphocytic choriomeningitis, 42
lysogeny, 61–7
lysosomes, 46, 48, 100, 101

measles, 42, 84, 111, 124, 127–8, 163
see also measles virus
measles virus, 6, 42, 77, 82, 103, 106
see also measles
membranes,
damage to, 92, 93, 94, 101, 103
entry via, 45–8
fusion of, 47, 48, 82
structure of, 29–31
uses of, 225–7
viral, 35, 38, 59–60, 225–7
meningitis, 42, 104, 111
Merck, Sharp and Dohme, 168
MMTV, 133
Montagnier, Luc, 193
mouth, *see* alimentary tract
mucus, 79, 81, 84, 96, 111, 112, 113, 116, 117

Mulligan, Richard, 220, 221
multiple sclerosis, 199–200
multiplication of viruses, 43–61, 64, 65, 69–71
mumps, 42, 77, 84, 111, 163, 167
muscles, 115, 116
mutations, 117, 147, 156, 163, 210, 211, 212, 213, 215
myxomatosis, 97–8

nasopharyngeal carcinoma, 139
natural killer cells, 86, 87, 89, 90
nervous system, 80, 84, 104, 113, 114, 125, 126, 128
  see also brain
neuraminidase, of influenza viruses, 117–19
nucleic acids, 14–20, 24–8, 72
  infectious, 203–6, 208–215
  "plus" and "minus", 53
  viral, 10, 12, 13, 34–41, 58
  see also DNA; genes; RNA
nucleosides, 15, 176, 177, 179
nucleotides, 14–20, 24–8, 54
  as drugs, 184–5
  see also nucleic acids

2′,5′-oligoadenylate (2,5-A), 183
orf virus, 42
origin of viruses, 206–16
orthomyxoviruses, *see* influenza viruses

pain, 104, 113, 115
pancreas, 84, 196–7
papillomaviruses, 42, 140
papovaviruses, 42, 65, 68
paramyxoviruses, 42
parvoviruses, 55, 56, 198
Pasteur, Louis, 1
peptide bonds, 21–2
peptides, 21, 170–2, 184
persistence of viruses, *see* infection, persistent
phagocytes, 86, 90, 91, 92, 93, 94, 114
phosphate, in nucleic acids, 15–17
picornaviruses, 42
placenta, 81
plasmids, 209, 210, 211
pneumonia, 116, 190
polioviruses, 6, 42, 58, 59, 78, 80, 81, 98, 99, 102, 163, 167, 172
"poly I:C", 183
polypeptides, *see* proteins
potato spindle tuber viroid, 205
poxviruses, 56–7, 104
  origin of, 213–15
  size, 7–9
  structure, 42
  see also smallpox
prions, *see* scrapie
prostaglandins, 114
proteins, 13, 20–8
  anti-viral, 88, 89
  "infectious", 201–4
  in membranes, 30–1, 225–7
  as receptors, 44–6, 79–81, 88, 89, 90, 91, 92, 93, 95, 96, 148, 149, 184
  synthesis of, *see* transcription; translation and vaccination, 168–74
  viral, 10, 12, 33–41, 45, 48, 49, 55–7, 58, 60, 64, 65, 211–12
  see also antibodies; antigens; enzymes; genetic engineering; glycoproteins; interferon; interleukin; vaccines

rabies virus, 6, 42, 78, 111, 163, 166, 181
radiation, 71, 120, 142, 155, 156
  see also, ultra-violet light receptor proteins, *see* proteins, as receptors
release of viruses from cells, 50, 58–60, 71
reoviruses, 42
replication,
  of DNA, 17–19, 101
  drug inhibition of, 176–9
  of RNA, 19
  of viral genes, 49, 50, 51, 54, 55
respiratory syncytial virus, 42
respiratory system, 37, 42, 75–8, 79, 81, 84, 96, 111, 112, 113, 116, 224
retroviruses, 54, 69–71, 184, 185
  and cancer, 133, 145–57
  in gene therapy, 220–5
  origin of, 211, 212

reverse transcriptase, 69, 70
rhabdoviruses, 42
rheumatoid arthritis, *see* arthritis, rheumatoid
rhinoviruses, 42, 79, 81, 180
ribonucleic acid, *see* RNA
ribosomes, 25–8
RNA, 14–20
  infectious, 204–6, 208–13
  in interferon induction, 88, 89, 183
  introns, 208, 209, 210, 211
  messenger, 25–8, 52–4, 56, 89, 208, 209, 210, 211
  replication of, 19
  of retroviruses, 69–71
  structure, 14–16
  transcription, 24–8, 52–4, *see also* transcription
  transfer, 26–7
  translation of, 24–8, *see also* translation
  as viral component, 35, 37, 39
Robinson, William, 137
Roche, 180
rotaviruses, 42
Rous, Peyton, 133, 146
rubella, 42, 124, 128–9, 163

saliva, 78, 84, 130
schizophrenia, 189
scrapie, 200–4
shingles, 42, 124, 126–7, 182
shivering, *see* fever
sickle cell anaemia, 220
Simons, Kai, 47–8
size of viruses, 7–9
skin, 76, 78, 80, 84, 126, 127, 190, 224
Slopek, Stephen, 218
smallpox, 42, 160, 162, 163, 164–7, 228
spleen, 84
SSPE, 124, 127–8
Stanley, Wendell, 33–5
Stephenson, Mary, 184
stomach, *see* alimentary tract
structure of viruses, 33–42
sugars, 31–32
  in nucleic acids, 15–17
SV40 virus, 65, 68

temperature, effect on viruses, 81
  see also fever
tetanus, 181
thalassaemia, 220
throat, *see* respiratory system; alimentary tract
thymidine phosphate, 15
thymine, 16, 17
tobacco mosaic virus,
  assembly of, 57–8
  discovery of, 1–5
  RNA of, 35, 36, 39
  structure of, 33–6
togaviruses, 42
transcription, 24–8, 52–4, 101
  of split genes, 209
  of viral genes, 49, 50, 51, 52–4, 55
transfer RNA, 26–7
transformation, 65, 68, 142–5
translation, 24–8
  of viral genes, 49, 50, 51, 52, 54, 55, 101
transmission of viruses, 75–9
transposons, 209–10, 211, 212, 213
tumours, *see* cancer
Twort, Frederick, 6–7

ultra-violet light, 63, 126, 163
uncoating of viruses, 49, 50
uracil, 16, 26
urogenital system, 76, 79, 81, 84, 123–6, 130

vaccines, 96, 128, 129, 159–74, 217, 220, 228
  against AIDS, 194
  against foot-and-mouth disease, 172
  against hepatitis B, 136, 138, 139, 167, 168–70, 172, 173
  against herpes, 167, 173
  against influenza, 163, 166, 167, 172, 173
  against measles, 163
  against mumps, 163, 167
  using peptides, 170–2
  against polio, 163, 167, 172
  against rabies, 163, 166
  against rubella, 163

against smallpox, 160, 162, 163, 164–7
against yellow fever, 167
see also immunisation
vaccinia virus, 42, 173
varicella-zoster virus, 42, 126, 127
see also chickenpox; shingles
variola virus, 42
see also smallpox
virinos, 204
viroids, 204–6, 210–11

warfare, *see* weapons
warts, 42, 78
water, 78
Waterfield, Michael, 147
Waterson, Anthony, 134
Watson, James, 19
weapons, 228–9

X-ray diffraction, 34–5

yeasts, in genetic engineering, 87, 168, 169, 170
yellow fever, 42, 78, 167
Yoon, Ji-Won, 196

Zamecnik, Paul, 184
Zimmern, David, 210–11
zinc, as an antiviral, 186–7